# Applications of Data Science to Molecules and Materials 2020



Topical Conference at the 2020 AIChE Annual Meeting

Online
16 – 20 November 2020

ISBN: 978-1-7138-2287-5

**Printed from e-media with permission by:**

Curran Associates, Inc.
57 Morehouse Lane
Red Hook, NY 12571



AIChE
120 Wall Street, FL 23
New York, NY 10005-4020

Phone: (800) 242-4363
Fax: (203) 775-5177

www.aiche.org

**Additional copies of this publication are available from:**

Curran Associates, Inc.
57 Morehouse Lane
Red Hook, NY 12571 USA
Phone: 845-758-0400
Fax: 845-758-2633
Email: curran@proceedings.com
Web: www.proceedings.com

# TABLE OF CONTENTS

# TABLE OF CONTENTS

# 599612 Discovery of Light-Driven Hydrogen Evolution Catalysts Using High Throughput Simulation, Experiments and Machine Learning

**Abstract:**

The renewable production of hydrogen is an important technology. Current approaches to catalyst discovery for thermo/electro and photochemical methods for hydrogen production have been difficult to optimize. This difficulty arises in part because of the separation of the synthesis and characterization of catalysts from their performance under operating conditions. It is challenging to build models that relate catalyst performance to synthesis conditions. We have taken a different approach to this problem. We use a photo-driven process that couples a photoabsorber with the synthesis of multi-component metal particles and hydrogen production. This approach is readily parallelized in a multi-well plate, enabling us to explore a wide range of synthesis conditions that are directly connected to catalyst performance measurements. We will show how we used this approach to screen thousands of possible catalyst synthesis conditions to identify promising combinations. The high-throughput experiments along, however, do not tell us what metals to test. For this we augment the approach with a parallel high-throughput computation approach where hundreds of thousands of DFT calculations are performed on multi-component alloy surfaces to screen for promising candidates with desired hydrogen and CO adsorption energies. Finally, we combine these efforts to drive focused studies of the most promising candidate synthesis conditions to develop a more conventional understanding of how the synthesis conditions led to the observed activity. We will show how this approach has led to confirmation of some known results from the electrochemical production of hydrogen from water, as well as some new results we may not have discovered with any single approach.

# 601035 High Throughput Screening of Alloy Structures for Propane Dehydrogenation Reaction

**Abstract:**

Non-oxidative propane dehydrogenation (PDH) on intermetallic alloys is a promising technology to convert large amounts of propane available in the shale gas to propylene. However, the high temperatures required for the reaction lead to reduced selectivity towards propylene formation and deactivation of the catalyst surface.[1] Even though Pt and Pd alloys have shown promise in terms of improved selectivity in comparison to pure metals, they ultimately experience loss of reactivity under reaction conditions. This difficulty motivates development of a more fundamental understanding of why Pt and Pd alloys have improved performance compared to pure metals and, further, elucidation of how such insights can be leveraged to find other alloy combinations with simultaneously high selectivity, stability, and activity for PDH.

Through our previous analysis on PdIn alloys, we have demonstrated that, as compared to terraces, the step surfaces of 1:1 alloys are highly active under PDH conditions (5 orders of magnitude higher rates). Furthermore, a comprehensive microkinetic analysis pointed to useful descriptors for activity, selectivity and stability.[2] With these descriptors as a starting point, the current study focuses on high-throughput screening of intermetallic alloy structures, including both steps and terraces, to identify reactivity trends across Pt and Pd alloys for PDH and, ultimately, find improved catalysts for this chemistry. First, we deploy a simple automated framework for adsorption and activation energy calculations on large number of chosen terrace and step alloy structures. The Materials Project is used to identify the most stable bulk structure of the alloy composition, while a surface energy analysis points to the most stable surface terminations, and CatKit is employed to identify all the distinct adsorption sites [3]. Subsequently, an in-house algorithm identifies all unique adsorbate configurations and finds the most stable adsorption site. These data are organized with a python-based databasing tool, leading to identification of BEP-type correlations between bond breaking barriers and the binding energies of various adsorbates. Further the binding energies of key descriptors were correlated with electronic and geometric properties of the alloy surfaces, using Random Forest techniques. This, therefore, allowed for a deeper understanding of the surface features that impacts the binding of descriptors the most.

**References:**

[1] J.J.H.B. Sattler, J. Ruiz-Martinez, E. Santillan-Jimenez, B.M. Weckhuysen, Catalytic dehydrogenation of light alkanes on metals and metal oxides, Chem. Rev. 114 (2014) 10613–10653. https://doi.org/10.1021/cr5002436.

[2] R.R. Seemakurthi, et al. Structure Sensitivity and Microkinetic Analysis of Propane Dehydrogenation on PdIn alloy and Pd, In Preparation.

[3] J.R. Boes, O. Mamun, K. Winther, T. Bligaard, Graph Theory Approach to High-Throughput Surface Adsorption Structure Generation, J. Phys. Chem. A. 123 (2019) 2281–2285. https://doi.org/10.1021/acs.jpca.9b00311.

# 609929 Theory-Guided, Interpretable Machine Learning Finds Predictive Geometric Structure-Property Relationships for Chemisorption on Alloys

**Abstract:**

Chemisorption, the direct chemical bonding of an atomic or molecular species to a solid-state material, is central in the fields of catalysis, corrosion, and electrochemistry, among many others. Despite the fundamental nature and importance of chemisorption, linking the geometry (i.e., the structure and composition) of different materials to their chemisorption properties remains a critical challenge. Developing physically transparent and quantitatively accurate models that can relate the chemisorption strength between an adsorbate and a solid surface to the adsorption site's geometry is critical to advance our understanding of chemisorption. In this talk, we discuss our efforts to use a theory-guided machine learning approach, which uses an interpretable class of machine learning models called generalized additive models (iGAM models),[1] to discover predictive structure-property models that can quantify the chemisorption strength of O, OH, S, and Cl on Pt-metal and Au-metal alloy surfaces subject to various strain- and ligand-induced changes in the local geometric structure. The iGAM models show a strong degree of predictive accuracy, with an average root-mean-square-error (RMSE) of 0.046 eV for samples in the test set. Through quantification of the relative importance of the features used to construct the models, we identified three important geometric features of the adsorption site that impact the relative chemisorption strength on metal alloys: the strain in the surface layer, the number of *d*-electrons in the ligand metal, and the size of the ligand atom. The interpretable functional form of the iGAM models allows us to analyze the model behavior with respect to each of these geometric features. Comparison between the iGAM chemisorption models and established electronic-structure models shed light on the critical physical concepts that control the chemisorption process on metal surfaces.

**References**

(1) Lou, Y.; Caruana, R.; Gehrke, J. Intelligible Models for Classification and Regression. In *Proceedings of the 18th ACM SIGKDD International Conference on Knowledge Discovery and Data Mining*; KDD '12; ACM: New York, NY, USA, 2012; pp 150–158. https://doi.org/10.1145/2339530.2339556.

# 608786 Trends in Catalyst Stability and Reactivity: Extracting Physical Insights Using Simple Data-Driven Approaches


**Abstract:**

The catalytic turnovers of a typical reaction pathway are determined by the energies of between 10 to 100 reaction intermediates.[1] Discovering optimally performing catalysts entails maximizing catalytic turnovers across vast spaces of energy and materials. One way to simplify this formidable optimization problem is to parametrize the reaction mechanism using energy correlations.[2] These energy correlations, often linear functions, are mappings between binding energies of reaction intermediates (e.g. $C_xH_yO_z$*) and one to two catalytic descriptors (e.g. binding energies of CH* and O*). Such energy correlations have enabled the high-throughput screening of metals, oxides, and zeolites. The slopes of these linear correlations (e.g. 0.5 for correlations between OH* and O*) across metals, oxides, and zeolites are determined by the ratios of bond orders of the corresponding species. Recent advances notwithstanding, these correlations are limited by three key features. First, since these correlations do not include catalyst stability metrics, they are unable to describe catalyst dynamics under reaction conditions. Second, within the existing formalism it is difficult to reverse-engineer active sites through property (e.g. rate) → structure (e.g. chemical composition) "inverse design" approaches. Third, their linearity, together with rigid constraints on the slope, imposes severe limitations in the accessible regions of materials space. There has been a concerted effort in the computational catalysis community to bridge these three knowledge gaps using data-driven statistical learning models.[3]-[5] Herein, we illustrate our contributions towards reverse-engineering realistic active ensembles sites that possess targeted catalytic turnovers.[6]-[9] We place a special emphasis on extracting physical insights from regressed parameters within our simple data-driven approach.[6], [10]

In the first part, we present accelerated schemes for determining the energies of metal atoms that, in turn, influence catalyst stability.[6] Predicting energies of metal atoms with atomic site specificity enables expeditious evaluations of surface energies, cohesive energies of nanoparticles having generic shapes, and binding energies of metal adsorption sites. We calculate the energies of metal atoms by melding quadratic and linear interpolations across the space of coordination numbers, composition, and external strain.[7] These new interpolations shrink the training set size for 12 fcc p- and d-block metals from > 204 to as few as 24 Density Functional Theory (DFT) calculations, without sacrificing accuracy on the test set.[6] We will specifically highlight physical insights regarding metal-metal bonding that are obtained from the regressed parameters. As a proof of concept, we employ this method to reverse engineer thermodynamically stable active site motifs of nanoparticles that catalyze $C_1$ conversion.

In the second part, we investigate the existence of energy correlations on supported nanoparticles using Au nanorods on doped MgO supports as a prototypical model.[10] Using > 1700 DFT calculated binding energies, we show that although DFT-derived energy correlations between 31 reaction intermediates of CO oxidation, water gas shift, and methanol synthesis do exist, their slopes depart from bond order conservation constraints. This departure from bond order conservation at bifunctional metal/oxide interfaces enhances the flexibility in materials space during catalyst optimization. We build a conceptual framework using the multipole expansion in electrostatics to semi-quantitatively explain trends within the slopes. Through these trends, we elucidate the electrostatic origin of the departure in slopes from bond order conservation.

Taken together, these two studies reveal new functional forms for energy correlations that determine metal nanoparticle stabilities, and reactivity trends at bifunctional interfaces. More crucially, we show that parameters of simple data-driven models contain a wealth of intriguing physical insights that enhance model interpretability.

**References**


1. Norskov et al., *Proc. Natal. Acad. Sci. USA,* (2011), **108,** 937.
2. Abild-Pedersen et al., *Phys. Rev. Lett.* (2007), **99,** 016105.
3. Tran et al., *Nature Catal.* (2018), **1**, 696.
4. O'Connor et al., *Nature Catal.* (2018), **1**, 531.
5. Goldsmith et al., *AICHE J.* (2018), **64,** 2311.
6. Choksi, Streibel, and Abild-Pedersen., *J. Chem. Phys.* (2020), **152,** 094702.
7. Streibel, Choksi and Abild-Pedersen. *J. Chem. Phys.* (2020), **152,** 094701.
8. Choksi et al. *J. Phys. Chem. Lett.* (2019), **10,** 1852.
9. Roling, Choksi, and Abild-Pedersen. *Nanoscale* (2019), **11,** 4438.
10. Choksi et al. *Angew. Chem. Int. Ed.* (2018), **57,** 15410.

# 604578 Orienteering in an Uncharted Chemical Space: Searching for an Optimal Bimetallic Nanocatalyst

**Abstract:**

Nanoparticles (NPs) have received tremendous attention as catalysts due to their high specific surface area and abundance of unique active sites. When designing such a catalyst, the Sabatier principle tells us that it must be tuned to adsorb key intermediates neither too strongly nor too weakly. For this reason, it is imperative to screen candidate NPs for their adsorption behavior. This task is conventionally accomplished either through time-consuming experiments, or computationally-expensive *ab-initio* calculations, which are primarily limited to highly idealized periodic surfaces. This presents a bottleneck in materials discovery, one which can be overcome through the development of rapid methods to predict adsorption behavior. To this end, we present a universal model of small molecule adsorption to NPs [1], which is able to accurately capture the adsorption behavior of metal NPs of any size, shape, or composition with minimal computational cost. By combining this adsorption model with our new genetic algorithm approach to NP structure prediction, we are now able to efficiently traverse the vast chemical space of bimetallic NPs and identify possible catalytic NPs to feed experiments. Overall, this work leverages cutting-edge chemical modeling techniques, such as first principles calculations, machine learning and statistical regression to significantly accelerate materials discovery and enable the rapid identification of promising bimetallic nanocatalysts.

1. Dean, J.; Taylor, M.; Mpourmpakis, G. Unfolding Adsorption on Metal Nanoparticles: Connecting Stability with Catalysis. *Science Advances* 5 (9), eeax5101.

# 604822 Physics Informed Machine Learning of Chemisorption at Metal Surfaces

**Abstract:**

The formation and breakage of chemical bonds at active sites is the molecular basis of catalysis. Being able to rapidly compute interaction strengths between bonding entities and understand their trends holds the key to the design of improved catalysts. Despite recent advances, machine learning (ML) faces a tremendous challenge for catalysis applications due to its poor transferability and explainability. Here we present a physics informed machine learning (PIML) approach that integrates convolutional neural networks with the *d*-band theory of chemisorption for predicting the chemical reactivity of metal surfaces. With *OH and *CO as two representative adsorbates, we demonstrated that the hybrid ML models outperform the purely data-driven ones in both data scarce and rich regions, especially for out-of-sample systems. More importantly, the architecture design enables its physical interpretability, shedding light on the nature of chemical bonding at metal surfaces.

# 606955 Improved Catalyst Predictions with Machine Learning Coupled Alchemical Perturbation Density Functional Theory

**Abstract:**

First principles Kohn-Sham density functional theory (KS-DFT) calculations hinder large-scale searches for new and sustainable heterogeneous catalysts due to significant computational expense. Alchemical perturbation density functional theory (APDFT) is a promising method that provides rapid predictions of catalyst descriptors based on perturbations of electrostatic potentials that arise from isoelectronic transmutations made to a reference catalyst. While straightforward and simple, first-order APDFT approximations are largely inaccurate when large and/or many transmutations are made. Here, we first discuss systematic trends in errors related to these factors for binding energy (BE) prediction of $CH_x$, $NH_x$, and $OH_x$ adsorbates on hypothetical alloy variations of fcc Pt(111) surfaces. Based on these observations, we demonstrate our approach fingerprinting locations of transmutations in these hypothetical alloys, and we show how to correct APDFT errors with machine learning models trained using about 3600 BE data points. Our models provide improved prediction accuracy with errors decreased by as much as an order of magnitude, and this approach extends the breadth of APDFT predictions, allowing rapid and accurate BE predictions on many more alloys made by numerous transmutation combinations.

# 597813 Molecular Design Blueprints: Catalysts and Principles from New Simulation and Machine Learning Tools

**Abstract:**

Many highly selective catalysts have been discovered that are defined by well-isolated sites with tailored metal-organic bonding. The rational design of such sites in de novo transition metal complexes, metal-organic frameworks (MOFs), or single atom catalysts, however, remains challenging. First-principles (i.e., with density functional theory, or DFT) high-throughput screening is a promising approach but is hampered by high computational cost, particularly in the brute force screening of large numbers of materials. In this talk, I will outline our efforts over the past few years to accelerate the design of single-site catalysts. I will describe our software and machine learning (ML) models that both simplify and accelerate the screening of new materials. These tools have enabled us to uncover new design rules and exceptions in both molecular and periodic materials (e.g., MOFs). We have paired ML models with multiobjective optimization strategies, robust uncertainty quantification, and highly parallel accelerated computing to discover optimal materials in weeks instead of decades. Time permitting, I will describe our recent efforts in autonomous computational chemistry by developing ML models as decision engines capable of predicting when calculations will fail and when methods beyond DFT are needed.

# 600080 Machine Learning Quantum Tunneling in the Kinetics of Chemical Reactions

**Abstract:**

The computation of reaction rate constants has been tackled using a multitude of theories ranging from classical transition state theory[1] to the fully quantum flux-flux correlation function approach.[2] The main cost of these methods is related to the need to explore potential energy surfaces, PES, either at the level of geometry optimizations, minimum energy path searches or to describe dynamics in time. In this context, machine learning algorithms which accelerate the search for reactive pathways or the prediction of rate constants are of great interest.[3,4] We have constructed a database of over 130000 reaction rate constants[5] obtained via quantum calculations of the reaction rate constants for particles crossing one dimensional single and double symmetric and asymmetric barriers. A deep neural network, DNN, was trained on the database with the ADAM optimizer and used to predict reaction rate constants. The most significant input features were identified using the Pearson correlation coefficient, and a grid search analysis was carried out for a coarse optimization of the DNN model hyperparameters. The network was trained for hundreds of epochs. The predicted results from the test set are in good agreement with the exact values. This approach shows promise for the application of machine learning to predict reaction rate constants.

1. Pechukas, P. Transition State Theory. *Annu. Rev. Phys. Chem.* **32**, 159–177 (1981).
2. Tromp, J. W. & Miller, W. H. The reactive flux correlation function for collinear reactions H + H2, Cl + HCl and F + H2. *Faraday Discuss. Chem. Soc.* **84**, 441–453 (1987).
3. Amabilino, S. *et al.* Training neural nets to learn reactive potential energy surfaces using interactive quantum chemistry in Virtual Reality. *J. Phys. Chem. A* **123**, 4486–4499 (2019).
4. Ulissi, Z. W. *et al.* Machine-learning methods enable exhaustive searches for active Bimetallic facets and reveal active site motifs for CO2 reduction. *ACS Catal.* **7**, 6600–6608 (2017).
5. Valleau, S. Machine learning quantum tunneling in the kinetics of chemical reactions. *In preparation.* (2020).

# 611092 Improved Catalyst Design and Performance through Application of Machine Learning Algorithms


**Abstract:**

Conversion of biomass feedstocks to hydrocarbon fuels is often fraught with catalyst deactivation due to deposition of coke on catalyst surface or deactivation due to inorganic components of the biomass. Acidic zeolite catalyst get easily deactivated because of the basic oxides in the biomass feedstock. The high aromatic content and macromolecules in the biomass pyrolysis oils also tend to promote carbon formation on the active sites, which eventually lead to their deactivation. We developed a nickel supported on mixed oxide (Ni/RM) catalyst that is more robust than the commercial nickel on silica-alumina catalyst. This new catalyst is regenerable and can be used effectively for hydrodeoxygenation of biomass pyrolysis oils to long chain hydrocarbon fuels. The new catalyst is effective in opening both furan and benzene rings and thus producing hydrocarbon fuels with linear and branched carbon chain length between C6 to C15. The catalyst was not deactivated by inorganic components in the biomass, however, the catalyst activity decreased with deposition of coke on the active sites and required frequent catalyst regeneration through combustion and reduction to restore activity. We developed a machine learning (ML) algorithm which enabled us to reformulate the catalyst and reduce the carbon deposition on the active site. The *signature* of each nickel supported mixed oxide catalyst (Ni/RM)) was defined as the weight percent of *eleven elements* (Al, Ca, Fe, K, Mg, Na, Ni, P, S, Si, Ti). These together with the *reaction temperature formed the feature vector* (input to ML). The output of the ML was coke formation, which was also a performance measure for the designed catalyst. The network relating the ML inputs to the output was either a simple Neural Network or a Convolutional Neural Network (CNN). Using this ML model as a basis, we inverted the model (inverse modeling) to obtain a new Ni/RM catalyst, which was then validated through experiments. The machine learning approached reduced catalyst development time and improved the robustness of the catalyst. Thus, the carbon deposition rate was reduced and catalyst regeneration frequency was also reduced.

# 603620 Improving the Reusability of Machine Learning Models for Materials

**Abstract:**

Machine learning models have proven useful in a variety of chemical and materials applications. However, one drawback of these models is that they are often geared to a particular application, and thus a new model must be created for a new application. This is particularly noticeable in applications that feature guest-host interactions, broadly defined, as there is a huge combinatorial challenge in considering many possible guests and many possible hosts. For example, this combinatorial challenge applies to catalysis, adsorption, doping, and intercalation. In this work, we develop and apply strategies to improve the generality and reusability of machine learning models.

We have developed a machine-learning architecture that takes advantage of inherent aspects of chemical systems, and is particularly geared towards those that feature guest-host interactions. This framework partially decouples the guests from the hosts by introducing intermediate variables, which are a set of implicit host properties that control how the host interacts with all guests. These intermediate variables are learned during the fitting process. We create separate sub-models for different host elements for predicting these intermediate variables, and separate sub-models for how different guests respond to these host variables. The sub-models are all fit simultaneously. Our framework takes advantage of the fact that elements are discrete entities, and greatly simplifies the huge combinatorial challenge of considering many possible guests and many possible hosts. We apply this framework to multiple datasets and show how it can be used to improve the efficiency and accessibility of initial screening in a variety of applications.

# 610067 The Open Catalyst Project Dataset

**Abstract:**

The Open Catalyst Project aims to develop new ML methods and models to accelerate the catalyst simulation process for renewable energy technologies and improve our ability to predict activity/selectivity across catalyst composition. To achieve that in the short term we need participation from the ML community in solving key challenges in catalysis. One path to interaction is the development of grand challenge datasets that are representative of common challenges in catalysis, large enough to excite the ML community, and large enough to take advantage of and encourage advances in deep learning models. Similar datasets have had a large impact in small molecule drug discovery, organic photovoltaics, and inorganic crystal structure prediction. We present the first open dataset from this effort on thermochemical intermediates across stable multi-metallic and p-block doped surfaces. This dataset includes full-accuracy DFT calculations across 53 elements and their binary/ternary materials, various low-index facets. Adsorbates span 56 common reaction intermediates with relevance to carbon, oxygen, and nitrogen thermal and electrochemical reactions. Off-equilibrium structures are also generated and included to aid in machine learning force field design and fitting. Collectively, this dataset represents the largest systematic dataset that bridges organic and inorganic chemistry and will enable a new generation of catalyst structure/property relationships. Fixed train/test splits that represent common chemical challenges and an open challenge website will be discussed to encourage competition and buy-in from the ML community.

# 607970 Active-Learning Framework for the Discovery of Novel Materials for Catalysis


**Abstract:**

The discovery of abundant and affordable materials for sustainable energy applications is a significant challenge of this decade. The space of possible material compositions and crystal structures is so vast that brute-force screening approaches are unfeasible. However, density functional theory (DFT) combined with machine-learning (ML) techniques have a huge potential to speed up the search and hold a great promise for the discovery of novel materials.

Here, we present a ML-based active-learning framework to search for stable and meta-stable inorganic materials. The workflow includes the generation of candidate structures, which is followed by automated DFT job submission, ML-training, and acquisition. When using a search space of experimentally observed crystal structures, this approach was recently applied to the discovery of new stable polymorphs of $IrO_x$, predicted to have an enhanced stability and catalytic activity for the oxygen evolution reaction (OER). In this talk, we extend the search to hypothetical crystal structures that are generated from a bottom-up prototype enumeration scheme, based on spacegroup and Wyckoff positions. This is applied to the search for cheaper and more abundant materials. Last, we envisioned how the screening approach can be extended to surface and chemisorption properties, on order to extend the search to catalytic applications.


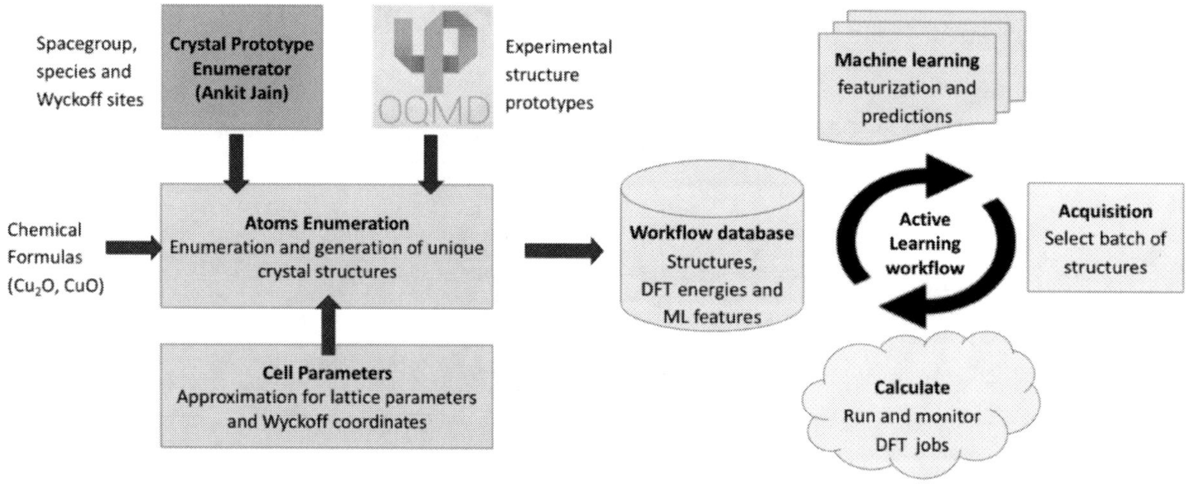

# 606321 Identifying Interaction Trends between Single Metal Atoms and Modified MgO(100) Supports with Statistical Learning

**Abstract:**

In this work, we apply statistical learning (SL) to identify physical descriptors for predicting electronic metal-support interactions (EMSI) in catalysis. Charge transfer between metal atoms and their underlying oxide support influences both the binding strength of the metal to the support and the binding strength of reaction intermediates to the metal. Tuning charge transfer by introducing surface-modifying dopants or co-adsorbates is an important strategy for controlling catalytic behavior. We use SL to build models for predicting metal atom binding energy to the oxide surface as a function of readily available physical descriptors containing properties of the metal, the support, and the surface-modifiers. Models are trained against DFT datasets consisting of metal adsorption energies on modified MgO(100) surfaces featuring various aliovalent dopants (e.g., Al, B, Li, and Na) and adsorbates (e.g., F, H, OH, and $NO_2$). These modifications generate both electron-rich and electron-poor MgO surfaces, which in turn enhance the extent of charge transfer between the metal and the support. We also test multiple SL procedures, including LASSO, Horseshoe prior, and Dirichlet-Laplace prior, and find that features derived by Dirichlet-Laplace prior construct the most accurate predictive models (i.e., achieving RMSE of ~0.2 eV for predicted binding energies). Moreover, we find that the features selected by SL using MgO training data generally are transferable to similar oxides, such as CaO(100), BaO(100), and ZnO(100), where data for these oxides are excluded from the feature selection process. This demonstrates the robust nature of features selected by the Dirichlet-Laplace prior procedure, both in terms of model accuracy and transferability to oxide surfaces beyond those included in the training set.

# 611372 How to Fix Some of Our Errors Via Data Science

**Abstract:**

Over the past two decades, multiscale modeling has advanced tremendously, and several algorithms currently exist. Yet, our ability to apply first principles modeling to process design is seriously limited due to multiple challenges. One of the challenges relates to error of first principles calculations, reaction mechanisms, active sites, etc. and its impact on catalyst discovery and reactor design. In this talk we will discuss sources of errors and new statistical learning methods that are transferable and explainable to quantify and correct some of the errors. We will discuss data quality and probabilistic graph models as a framework that integrates data, knowledge, and physical models to enable uncertainty quantification.

# 602185 Finding Optimal Reaction Routes in Chemical Reaction Networks


**Abstract:**

Large chemical databases, such as Reaxys®[1], containing more than 119 Million compounds and more than 46 Million reactions, have previously been used to investigate the connectedness of chemical space.[1] Big data sets of known chemical reactions have been explored as the Network of Organic Chemistry by representing molecules as nodes and reactions as edges.[2] Amongst others, the networks statistics,[1] the networks historic evolution,[3] and its usefulness for reaction development[4] and retrosynthesis[5] have recently been recognised.

Many questions about the chemical space and its applications, however, remain unsolved, e.g. the automated reaction route selection. That is choosing the *best* route from certain feedstock via multiple intermediates to desired end-products, whereby *best*still has to be defined. The problem has received much attention at present with regard to the growing need for more sustainable reaction routes and routes developed on novel feedstocks, i.e. renewable feedstocks and waste streams.[4] Optimisation based methods, e.g. the Reaction Network Flux Analysis (RNFA), have been recognised to enable route selection with manually mined and relatively small data sets (e.g. 80 substances and 116 reactions)[6]. It was firstly introduced by Voll and Marquardt for the evaluation of reaction routes for biofuels[6] and has been further developed for instance to extend the route evaluation criteria,[7] and to account for geographic locations and supply chain considerations.[8] Furthermore, Zhang et al. had implemented the RNFA in order to evaluate the reaction pathways for the production of biopolymers.[9]

Herein, automatable methods to detect optimal reaction routes from very large sets of recorded reactions are discussed. Firstly, we investigate two problem formulations for automatically mined chemical data sets.[10] In addition to the RNFA, we introduce a Petri Net Optimisation (PNO) formalism which explicitly takes the order in which reactions occur into account. In network optimisation this can enable constraints on reaction cycles and intrinsically enables us to regard non-flux related reaction costs. Secondly, we explore chemical heuristics and sustainability metrics derivable from digital data recording. We trim down data recordings using heuristics, e.g. variation of carbon atoms and exclusion of pre-defined chemical structures. Balance equations are formulated as constraints and a variety of sustainability metrics is explored as objective functions. We demonstrated the potential of the proposed method on an illustrative working example and a relevant chemical case study for sustainable production of citral from -pinene. Citral is an important compound in the fragrance and flavour industry and also an intermediate for pharmaceutical processes;[11] -pinene is has been shown as a valid starting material due to its structure, its occurrence in non-edible plants (pine trees)[12] and as co-product from the paper industry.[13]

This work bundles present knowledge and elucidates a roadmap for chemical process route selection as a driver for the transition towards a sustainable chemical industry.



**Acknowledgments:** J. M. Weber would like to thank the Department of Chemical Engineering for funding in the form of a PhD studentship and Philipp-M. Jacobs for his support and expertise. We gratefully acknowledge collaboration with RELX Intellectual Properties SA and their technical support, which enabled us to mine Reaxys. Reaxys data were made accessible to our research project via the Elsevier R&D Collaboration Network. This work was funded, in part, by the EPSRC project "Terpene-based manufacturing for sustainable chemical feedstocks" EP/K014889. This project was funded, in part, by the National Research Foundation (NRF), Prime Minister's Office, Singapore under its Campus for Research Excellence and Technological Enterprise (CREATE) program as a part of the Cambridge Centre for Advanced Research and Education in Singapore Ltd (CARES).

**References:**

1. Jacob, P. M. & Lapkin, A. A. Statistics of the network of organic chemistry. *React. Chem. Eng.* **3**, 102–118 (2018).
2. Bishop, K. J. M., Klajn, R. & Grzybowski, B. A. The core and most useful molecules in organic chemistry. *Angew. Chemie - Int. Ed.* **45**, 5348–5354 (2006).
3. Llanos, E. J. *et al.* Exploration of the chemical space and its three historical regimes. *Proc. Natl. Acad. Sci.* **116**, 12660–12665 (2019).
4. Jacob, P. M., Yamin, P., Perez-Storey, C., Hopgood, M. & Lapkin, A. A. Towards automation of chemical process route selection based on data mining. *Green Chem.* **19**, 140–152 (2017).
5. Szymkuć, S. *et al.* Computer-Assisted Synthetic Planning: The End of the Beginning. *Angew. Chemie - Int. Ed.* **55**, 5904–5937 (2016).
6. Voll, A. & Marquardt, W. Reaction Network Flux Analysis: Optimization-Based Evaluation of Reaction Pathways for Biorenewables Processing. *AIChE J.* **58**, 1788–1801 (2012).
7. Ulonska, K., Voll, A. & Marquardt, W. Screening Pathways for the Production of Next Generation Biofuels. *Energy and Fuels* **30**, 445–456 (2016).
8. Ulonska, K., König, A., Klatt, M., Mitsos, A. & Viell, J. Optimization of Multiproduct Biorefinery Processes under Consideration of Biomass Supply Chain Management and Market Developments. *Ind. Eng. Chem. Res.* **57**, 6980–6991 (2018).
9. Zhang, D., Del Rio-Chanona, E. A. & Shah, N. Screening Synthesis Pathways for Biomass-Derived Sustainable Polymer Production. *ACS Sustain. Chem. Eng.* **5**, 4388–4398 (2017).
10. Weber, J. M., Schweidtmann, A. M., Nolasco, E. & Lapkin, A. A. Modelling Circular Structures in Reaction Networks: Petri Nets and Reaction Network Flux Analysis. in *in Press: European Symposium on Computer Aided Process Engineering* (2020).
11. Maswal, M. & Dar, A. A. Formulation challenges in encapsulation and delivery of citral for improved food quality. *Food Hydrocoll.* **37**, 182–195 (2014).
12. Vespermann, K. A. C. *et al.* Biotransformation of α- and β-pinene into flavor compounds. *Appl. Microbiol. Biotechnol.* **101**, 1805–1817 (2017).
13. Winnacker, M. Pinenes: Abundant and Renewable Building Blocks for a Variety of Sustainable Polymers. *Angew. Chemie - Int. Ed.* **57**, 14362–14371 (2018).

# 605510 *Kinexns (Kinetics of Reactions)*: an Open Source Python Package for Chemical Kinetic Modeling

**Abstract:**

Chemical kinetic models are usually composed of a set of stiff ordinary differential equations (ODEs), where a number of rate parameters have to be estimated to fit with experimental observations. This presents a twofold challenge—firstly, to solve the stiff set of ODEs accurately and efficiently; and secondly, to estimate the model parameters precisely by optimizing the objective function. In recent years, stochastic optimization methods for parameter estimation have gained popularity over the classical optimization methods as the former do not require a reasonable initial guess and have the capability to escape local minima. Therefore, in this study, we develop an open source python package *kinexns* to efficiently solve chemical kinetic model using CVode stiff ODE solver, perform sensitivity analysis to determine important model parameters, and optimize the model parameters by using the different stochastic algorithms. We also systematically examine ten different stochastic optimization algorithms and evaluate their performance to estimate the model parameters for previously developed propane oxidation mechanism—popularly known as the *San Diego mechanism*.[1] The different algorithms we considered are: Monte Carlo (MC), Latin-Hypercube Sampling (*LHS*), Maximum Likelihood Estimation (*MLE*), Markov-Chain Monte-Carlo (*MCMC*), Shuffled Complex Evolution Algorithm (*SCE-UA*), Simulated Annealing (*SA*), RObust Parameter Estimation (*ROPE*), Artificial Bee Colony (*ABC*), Fitness Scaled Chaotic Artificial Bee Colony (*FSCABC*), and Dynamically Dimensioned Search algorithm (*DDS*). Our results indicate the MLE and DDS provide better parameter approximation among all the algorithms evaluated. This package is finally used to develop a robust kinetic model for fast pyrolysis of xylose by identifying important reactions through sensitivity analysis and optimizing the associated rate parameters.

(1) Chemical Mechanism: Combustion Research Group at UC San Diego https://web.eng.ucsd.edu/mae/groups/combustion/mechanism.html.

# 599788 Artificial Intelligence with Microreactors for the Activity and Mechanism of a Zirconocene-Catalyzed Olefin Polymerization

**Abstract:**

The design of artificially intelligent micro-scale reactors has the potential to revolutionize catalysis. When combined with *in situ* spectroscopy, they can generate high-fidelity, transient information faster, safer, and with less environmental impacts compared to conventional methods. This seminar will highlight our recent findings that motivate the use of artificial neural networks (ANNs) for a supervised, yet more thorough investigation of catalytic reaction kinetics and mechanisms [1-3]. In the first example, the temperature control of microfluidics by computer-vision based on ANNs will be introduced. Their application can accelerate, in the second example, the discovery of metallocene catalysts for highly complex and notoriously difficult to control exothermic olefin polymerizations when limited experimental kinetic data are available. Finally, ANNs combined with experimental automation can also help to validate such complex polymerization mechanisms or to discover the optimal catalytic activity. The future of supervised machine learning in chemical reaction engineering is promising, offering new methods to better understand catalytic reactions.

[1] Rizkin, B.A., Shkolnik, A.S., Ferraro, N.J., Hartman, R.L. *Nat. Mach. Intell.* (2020) DOI: 10.1038/s42256-020-0166-5.

[2] Rizkin, B.A., Hartman, R.L. *Chem. Eng. Sci.*, 210, 115224 (2019).

[3] Rizkin, B.A., Popovich, K., Hartman, R.L. *Comput. Chem. Eng.*, 121(2), 584-593 (2019).

# 610074 Meso-Kinetic Transient Optimization Via Machine Learning

**Abstract:**

Understanding the set of elementary steps and kinetics in a given reaction is extremely valuable for making informed decisions about next-generation catalyst materials. Traditionally, this information has been collected through extensive physics-based simulation to construct first-principles models, and/or fitting kinetic models with or without underlying mechanisms to experimental data. However, due to the rough granularity of commonly used steady-state analysis techniques, it is common that multiple first-principle models fit the experimental results equally well. As such, it is often up to the domain experts to infer the kinetics based on chemical intuition. We propose to help bridge the gap between simulation and experiment through the combination of highly dense transient kinetic measurement and chemically-inspired machine learning techniques. Specifically, we leverage non-convex variable selection techniques within a framework constrained by a chemical mechanism to obtain a quasi-equilibrium expression of the reaction mechanism purely based on experimental data. This methodology, referred to as MEso-kinetic TRansient Optimization (METRO), is validated against simulated responses where true kinetic coefficients are known, and is shown to accurately estimate these parameters without any a priori information. Moreover, the method is applied to experimental data from well explored reactions such as carbon monoxide oxidation and ammonia decomposition.The METRO methodology is proposed as a compliment to bottom-up microkinetic modeling, where the top-down fitting can provide insight into the reaction mechanism and extract intrinsic kinetic parameters of elementary processes directly from experimental transient kinetic data.

1 $CO + Z \longrightarrow COZ$

2 $O_2 + Z \longrightarrow 2OZ$

3 $COZ + OZ \rightleftharpoons CO_2Z + Z$

4 $CO_2Z \longrightarrow CO_2 + Z$

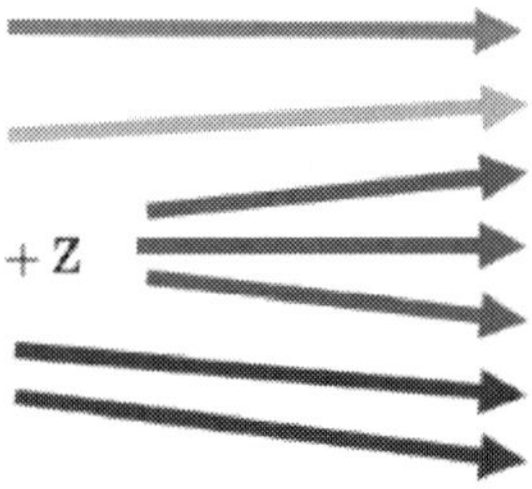

| | Expression | Coefficient | CI | Slope |
|---|---|---|---|---|
| 1 | $CO_R = CO_C$ | 63.516 | (53.929,73.103) | -0.169 |
| 2 | $O_{2R} = O_{2C}$ | -8.320 | (-8.823,-7.817) | NA |
| 3 | $CO_R = CO_C O_U$ | 6.531 | (5.799,7.264) | NA |
| 4 | $CO_{2R} = CO_{2C} O_U$ | 1.242 | (1.015,1.469) | NA |
| 5 | $O_{2R} = O_U$ | -1.955 | (-2.932,-0.977) | NA |
| 6 | $CO_{2R} = O_U$ | -0.024 | (-0.033,-0.016) | NA |
| 7 | $CO_{2R} = CO_{2C} CO_{2U}$ | -2.963 | (-3.709,-2.217) | NA |
| 8 | $CO_{2R} = CO_{2C}$ | 12.003 | (10.156,13.85) | -0.125 |

Figure 1. Comparison of the traditional form of the CO oxidation mechanism and the quantitative results obtained by the METRO algorithm. Each coefficient within the METRO algorithm describes an intrinsic rate coefficient while providing statistical bounds of the point estimation and the slope associated consumption of active sites over a series of experiments.

# 607498 Improving Microkinetic Models of Catalytic Reactions Using Machine Learning

**Abstract:**

Microkinetic modeling is a common tool in computational catalysis to predict or infer mechanistic details of catalytic systems. The state-of-the-art in the field is to develop ab initio microkinetic models where the kinetics and thermochemistry information is obtained from density functional theory (DFT) calculations. Several assumptions and approximations are often employed in this process, however, these lead to large predictive inaccuracies – (i) intrinsic errors from DFT propagate to errors in kinetic/thermodynamic parameters that are on the order of a factor of $10 - 10^3$; (ii) entropies are often calculated assuming the quantum harmonic oscillator approximation which results in significant errors in the calculation of free energies for intermediates with weak vibrational modes that are neither as free as translational/rotational modes nor as restricted as those corresponding to strong covalent bonds; (iii) rates are calculated using the mass-action kinetics and mean field concentrations of species on the surface.

In this talk, we show that machine learning techniques can be employed to address each of these three issues systematically to rigorously improve the predictive accuracy of microkinetic models while being computationally tractable. We will show that interesting challenges emerge here in the context of machine learning; in particular, data-driven models have to be built using "sparse" and "heterogeneous" data as opposed to "big" and "homogeneous".

Examples from transition metal catalysis will be used to discuss each of the proposed improvements.

# 596965 Characterizing Complex Solvent Environments in Acid-Catalyzed Reactions Using Molecular Dynamics Simulations and 3D Convolutional Neural Nets

**Abstract:**

The catalytic conversion of lignocellulosic biomass is a promising strategy to obtain transportation fuels and high-value chemicals from renewable feedstocks [1]. The conversion of biomass-derived molecules is typically facilitated by liquid-phase, acid-catalyzed reactions that are hindered by low reactivity in aqueous solution. One method to increase acid-catalyzed reaction rates is to modify the solvent composition by mixing organic, polar aprotic cosolvents with water to create mixed-solvent environments [2, 3]. Compared with trial-and-error experimentation, computational tools have been applied to understand solvent effects on chemical reactivity and guide the design of solvent mixtures for efficient and cheap biomass conversion processes [4]. Molecular dynamics (MD) simulations can be utilized to understand and predict solvent effects on experimental reaction rates for the conversion of biomass-derived model compounds in aqueous mixtures of 1,4-dioxane (DIO), g-valerolactone (GVL), and tetrahydrofuran (THF) [3]. We developed an MD model consisting of only reactant, water, and cosolvent molecules and calculated three simulation-derived descriptors for a linear regression model to predict experimental reaction rates and found good agreement in DIO-water mixtures [3]. The regression model was less accurate for GVL- and THF-water mixtures, indicating that either descriptor computed with classical MD cannot quantify reaction rates in these systems or that more complex descriptors must be defined to capture reactivity trends. However, designing new descriptors of reaction kinetics based on human intuition is challenging, often requiring complex and time-consuming data analysis tools (*e.g.* solvation free energies [6] or three-dimensional solvent mapping [5]) that cannot be readily generalized across a range of solvent compositions.

As an alternative to designing descriptors via human intuition, machine learning methods have been increasingly used to infer molecular properties by automatically extracting features from complex sources of data [7-13]. For example, convolutional neural networks (CNNs) can be used to identify and quantify patterns within two-dimensional (2D) spatial datasets such as images [14]. By training on a suitable set of labeled image data, CNNs extract spatial features without requiring human supervision and can then utilize these features to classify image contents. CNNs can be further generalized to extract features from three-dimensional (3D) volumetric data [15], which can facilitate the analysis of 3D molecular structures. For example, 3D CNNs have recently been used to detect protein functional sites [16], evaluate protein-ligand binding sites [17], and quantify protein-ligand binding affinities [18] by training on protein database structures. Based on these examples and our prior success using classical MD simulations to predict acid-catalyzed reaction outcomes [3], we hypothesize that 3D CNNs can exploit the output of classical MD simulations to more accurately predict solvent effects on acid-catalyzed reaction rates.

In this work, we developed 3D CNNs that utilize atomic positions obtained from classical MD simulation trajectories to predict the rates of liquid-phase, acid-catalyzed biomass conversion reactions in mixed-solvent environments. We constructed 3D grids of voxels (the 3D analogs of 2D pixels) that represent atomistic positions sampled in corresponding MD simulations. We find that our 3D CNN model, which we call SolventNet, predicts experimental reaction rates more accurately than models based on human-selected, MD-derived descriptors [3] and previously developed 3D CNNs (ORION [19] and VoxNet [20]). Surprisingly, reaction rate predictions with SolventNet require as little as 2 ns of classical MD trajectory data, a 100-fold improvement from the original 205 ns of MD data used in models based on human-selected descriptors [3]. This indicates that 3D atomistic positions embed significant information. We further show that SolventNet generalizes to new system compositions using leave-one-out cross-validation in which all data for a cosolvent-water mixture or reactant were treated as the test set and excluded from model training. Finally, we tested the predictive power of SolventNet for reactants in three additional polar aprotic cosolvents not included in model training: dimethyl sulfoxide, acetonitrile, and acetone. SolventNet still accurately predicts experimentally measured reaction rates in solvent mixtures containing these cosolvents despite their distinct properties (*e.g.*, functional groups, basicity, and polarizability). To our knowledge, this work is the first to integrate 3D CNNs and classical MD simulations for the prediction of acid-catalyzed reaction rates. We envision that the computational efficiency associated with the combination of 3D CNNs and classical MD simulations will enable the integration of these tools with process models to screen solvents and optimize reactor conditions for biomass conversion processes [21].

[1] L. Shuai and J. Luterbacher, *Chemsuschem*, 2016, **9**, 133-155.

[2] M. A. Mellmer, C. Sener, J. M. R. Gallo, J. S. Luterbacher, D. M. Alonso and J. A. Dumesic, *Angew Chem Int Edit*, 2014, **53**, 11872-11875.

[3] T. W. Walker, A. K. Chew, H. X. Li, B. Demir, Z. C. Zhang, G. W. Huber, R. C. Van Lehn and J. A. Dumesic, *Energy & Environmental Science*, 2018, **11**, 617-628.

[4] J. J. Varghese and S. H. Mushrif, *Reaction Chemistry & Engineering*, 2019, **4**, 165-206.

[5] S. H. Mushrif, S. Caratzoulas and D. G. Vlachos, *PCCP*, 2012, **14**, 2637-2644.

[6] A. K. Chew and R. C. Van Lehn, *Front Chem*, 2019, **7**, 439.

[7]Connor W. Coley, W. Jin, L. Rogers, T. F. Jamison, T. S. Jaakkola, W. H. Green, R. Barzilay and K. F. Jensen, *Chem Sci*, 2019, **10**, 370-377.

[8] D. K. Duvenaud, D. Maclaurin, J. Iparraguirre, R. Bombarell, T. Hirzel, A. Aspuru-Guzik and R. P. Adams, 2015.

[9] R. Gómez-Bombarelli, J. N. Wei, D. Duvenaud, J. M. Hernández-Lobato, B. Sánchez-Lengeling, D. Sheberla, J. Aguilera-Iparraguirre, T. D. Hirzel, R. P. Adams and A. Aspuru-Guzik, *ACS central science*, 2018, **4**, 268-276.

[10] N. E. Jackson, A. S. Bowen, L. W. Antony, M. A. Webb, V. Vishwanath and J. J. de Pablo, *Sci Adv*, 2019, **5**, eaav1190.

[11] E. Y. Lee, B. M. Fulan, G. C. L. Wong and A. L. Ferguson, *Proceedings of the National Academy of Sciences*, 2016, **113**, 13588-13593.

[12] Z. Wu, B. Ramsundar, E. N. Feinberg, J. Gomes, C. Geniesse, A. S. Pappu, K. Leswing and V. Pande, *Chem Sci*, 2018, **9**, 513-530.

[13] S. Chmiela, A. Tkatchenko, H. E. Sauceda, I. Poltavsky, K. T. Schütt and K.-R. Müller, *Sci Adv*, 2017, **3**, e1603015.

[14] W. Rawat and Z. H. Wang, *Neural Comput*, 2017, **29**, 2352-2449.

[15] R. D. Singh, A. Mittal and R. K. Bhatia, *Multimedia Tools and Applications*, 2019, **78**, 15951-15995.

[16] W. Torng and R. B. Altman, *Bioinformatics*, 2018, **35**, 1503-1512.

[17] J. Jiménez, S. Doerr, G. Martínez-Rosell, A. S. Rose and G. De Fabritiis, *Bioinformatics*, 2017, **33**, 3036-3042.

[18] J. Jiménez, M. Škalič, G. Martínez-Rosell and G. De Fabritiis, *J Chem Inf Model*, 2018, **58**, 287-296.

[19] N. Sedaghat, M. Zolfaghari, E. Amiri and T. Brox, *arXiv preprint arXiv:1604.03351*, 2016.

[20] D. Maturana and S. Scherer, *2015 IEEE/RSJ International Conference on Intelligent Robots and Systems (IROS)*, 2015, 922-928.

[21] D. M. Alonso, S. H. Hakim, S. Zhou, W. Won, O. Hosseinaei, J. Tao, V. Garcia-Negron, A. H. Motagamwala, M. A. Mellmer, K. Huang, C. J. Houtman, N. Labbé, D. P. Harper, C. T. Maravelias, T. Runge and J. A. Dumesic, *Sci Adv*, 2017, **3**, e1603301.

# 597607 A Data Driven Investigation of Catalytic Upgrading of Ethane to Ethylene over $Ga/Al_2O_3$

**Abstract:**

The recent shale gas revolution offers exciting opportunities to meet global demands with new solutions. Increasing demand for ethylene, one of the world's most produced chemicals, presents such an opportunity. While traditionally generated through naphtha cracking, ethylene may also be produced through the dehydrogenation of ethane present in shale gas. To enable the economic feasibility of this process, studies[1] screened numerous candidate catalysts and discovered alumina-supported gallium ($Ga/Al_2O_3$) as a highly promising material. However, its impressive catalytic activity is inhibited by deactivation due to coking. Attempting to prevent this deactivation while retaining dehydrogenation activity, multiple investigations[2,3] explored coking mitigation strategies, such as $CO_2$ co-feeding, but we have yet to understand the complete, underlying kinetics and unlock the potential of this exciting catalyst.

In this work, we investigate the kinetics of ethane dehydrogenation over $Ga/Al_2O_3$ and the associated deactivation. Analyzing our micro gas chromatography data with novel Bayesian techniques, we ascertain a layered reaction network and associated kinetic parameters. We discover a fundamental change stemming from the water produced that selectively affects ethane dehydrogenation but not the unwanted side reaction of hydrogenolysis. This water effect also applies to the deactivation kinetics and explains $CO_2$ co-feeding inhibiting coking through the surprising lens of the reverse water-gas shift reaction. Applying these findings alongside additional experiments exploring water co-feeding, we identify a promising path towards substantial conversion, improved selectivity, and negligible coking. These discoveries offer novel insights into this promising catalyst and general applicability to similar oxide materials.

1. Nakagawa, K., et al. Chemical Communications, no. 9 (1998): 1025–1026.
2. Xu, B.,et al. Journal of Catalysis, no. 2 (2006): 470–477.
3. Xiao, H., et al. Catal. Sci. Technol., no. 13 (2016): 5183–5195.

# 610831 Prediction of Adsorption Thermodynamics in MOFs and Cofs Via Ensemble Learning

**Abstract:**

We discuss how ensemble learning (EL) allows for a rapid prediction of the partition function for fluids adsorbed in nanoporous materials. For this purpose, we carry out flat histogram simulations for the adsorption process and train EL models on the simulated datasets. Once partition functions are predicted by the EL models, all thermodynamic properties of adsorption can be rapidly determined for a wide range of conditions, without any additional simulations. We apply various EL approaches to carbon capture and storage, hydrogen storage and gas separation in MOFs and COFs. In particular, we find that the diversity ensemble learning approach, obtained from neural networks with different types of architectures, provides the most accurate predictions, when compared to the experiment and to Expanded Wang-Landau simulation results. The EL models so obtained give access to the materials selectivity in the case of the dsorption of mixtures and to a number of key thermodynamic properties, such as the desorption free energy which measures the energetic cost to regenerate the adsorbent.

# 606275 Message Passing Neural Networks for Partial Charge Assignment in Metal-Organic Frameworks

**Abstract:**

Virtual screenings of thousands of metal-organic frameworks (MOFs) can accelerate and reduce the cost of discovering MOFs for their applications in gas storage, separation/purification, and sensing. In molecular mechanics simulations of gas adsorption/diffusion in MOFs, the adsorbate-MOF electrostatic interaction is typically modeled by placing partial charges on the atoms of the MOF. It is critical to develop computationally inexpensive methods to assign atomic partial charges to a MOF that accurately reproduce the electrostatic potential in its pores. Herein, we design and train a message passing neural network to predict the atomic partial charges on MOFs under a charge neutral constraint. A set of ca. 2266 MOFs labeled with high-fidelity partial charges, derived from electronic structure calculations, serve as training examples. In an end-to-end manner, from the crystal graph representing the MOF, our message passing neural network machine-learns features of the local bonding environments of each atom and learns to predict partial charges from these features. Our trained message passing neural network can assign high-fidelity partial point charges to MOFs with orders of magnitude lower computational cost, thereby enhancing the accuracy of virtual screenings of MOFs for their adsorption-based applications.

# 602469 High-Throughput Discovery of Metal-Organic Frameworks for Cooperative $CO_2$ Adsorption

**Abstract:**

Recently, a new class of post-synthetically modified metal-organic frameworks (MOFs) with non-Langmiur stepped isotherms have been discovered and tuned to reversibly and selectively adsorb CO2 under common flue gas conditions. However, very few MOFs are known to exhibit step-like isotherms, a result of a cooperative adsorption phenomenon. Here, we present a computational screening procedure to discover new CO2 adsorbent MOFs with the potential for step-like isotherms and cooperative adsorption. Our workflow is based on the hypothesis that the distance between accessible, undercoordinated metal sites is a key indicator for whether a MOF modified with ethylenediamine will exhibit cooperative adsorption. We screen the Computational-Ready Experimental MOF (CoRE-MOF) database using the fast marching algorithm to assess metal site distances given arbitrary pore geometries, and discuss candidate materials for experimental validation.

# 600720 Gaussian Process-Based Model Order Reduction for the Prediction of Gaseous Storage in Metal-Organic Frameworks.

**Abstract:**

Metal-organic frameworks (MOFs) are a unique class of crystalline porous materials with potential applications in a multitude of industrial settings including gas storage. The nature of MOFs means that their structures are highly tailorable allowing a large number to be synthesized. This poses the inherent problem of identifying promising MOFs for a given application without the need for complex and time-consuming calculations.

This study aims to further the understanding of MOFs by deriving a novel method that uses synthesised MOFs to develop a prediction technique that results in faster and more efficient material design and discovery for gas storage applications. The ability to accurately predict gas adsorption properties and identifying the most dominant geometrical features of MOFs affecting adsorption phenomena, allows scientists to quickly and efficiently design and discover new MOFs. To do this, we adopt a Gaussian Process (GP) surrogate modelling approach, applied to a large database of MOFs that have been experimentally synthesized. This GP surrogate is then used to perform a Global Sensitivity Analysis of the data, to quantify both the most important geometric characteristics of MOFs and the interaction between them, which is also then used as part of a dimensionality reduction approach to identify the parameter subspace required to accurately predict gas adsorption. A comparison is then made between these subspaces for the different adsorbed gases to assess the ability to predict across different gas species, here oxygen and methane.

Results, shown in Figure 1, demonstrate that the surface area and void fraction dominate the deliverable capacity of both oxygen and methane in MOFs. The dimensionality reduction shows that the pore properties can be confidently reduced to just three primary dimensions while also improving the prediction accuracy when the storage gas is varied. Overall, the use of a GP surrogate model for a Global Sensitivity Analysis and a reduced order model has revealed the accurate determination of important pore properties and has helped pave the way for accurate prediction of MOF performance regardless of the gas. This will aid experimental researchers in designing MOFs more efficiently concentrating on optimising the important pore properties.

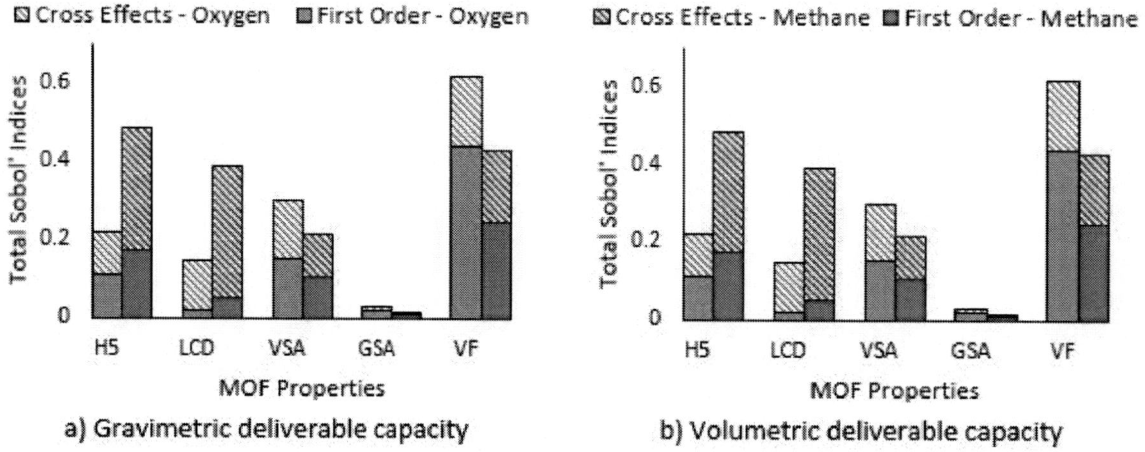


Figure 1: The bar charts showing the calculated total Sobol' indices for each geometric feature with respect to the deliverable capacity of MOFs.

# 596692 A Database with Automated Quantum Chemistry Calculations and Machine Learning for Functional Transition Metal Complex Discovery

**Abstract:**

Large-scale databases of density functional theory (DFT) properties have proven invaluable in accelerating materials discovery, with notable example databases in both the solid state and small organic molecules. A similar database of transition metal (TM) complexes with DFT properties, however, is absent due to unique challenges in structure curation and domain of application of DFT for their properties. To address these challenges, we have developed open-source toolkits molSimplify and AutomaticDesign that incorporate both chemical rules and data-driven machine-learning (ML) models. Here, we highlight the first large-scale computational database of TM complexes, containing over 40,000 attempts of DFT geometry relaxation of octahedral complexes. From this database, we select a subset of over 24,000 complexes with higher-quality geometry and electronic structure. For these complexes, we calculated properties such as solubility, spin-splitting energy, redox potential, frontier orbital energies, and energetic functional dependence. To extend the number of complexes in the database continuously over time, we further built a workflow to automatically perform DFT geometry relaxations on diversity search-selected experimentally-determined complexes (i.e., from the Cambridge Structure Database) when supercomputing resources are idle. We also describe our automatic ML-model retraining module, which updates all our ML models when new results become available. With this automated workflow, we can readily monitor the evolution and quality of our ML models as the database grows. We anticipate our database, enhanced by automated quantum chemistry calculations and ML model retraining, to be a useful tool in accelerating discovery of functional transition metal complexes.

# 598384 Do Machine-Learned Formation Energies Enable Accurate Predictions of Compound Stabilities?

**Abstract:**

Machine learning (ML) models in chemistry and materials science are often tasked with predicting compound formation energies. These formation energies can then be used as inputs to construct phase diagrams and determine the stability of materials. Recent efforts have shown that ML models trained on density functional theory (DFT) calculated formation energies for solid-state materials achieve accuracies comparable to DFT itself (relative to experiment). In principle, these ML models should therefore be able to predict the stability of materials with comparable accuracy to DFT at a miniscule fraction of the computational expense.

In this work, we designed a set of tests to assess whether seven recently published machine learning formation energy models can accurately predict the stability of compounds. These tests include the reconstruction of ~20,000 phase diagrams available in the Materials Project database in addition to a simulated materials discovery problem, where the models are tasked with re-discovering stable compounds in sparse chemical spaces. Our findings suggest that ML models based only on chemical composition are generally not capable of reliably predicting compound stability and sheds insights into why that is. We do show that stability predictions can be improved by the incorporation of structure in the material representation.

In total, this work emphasizes the importance of testing machine learning models on real-world applications to assess their potential for incorporation in the materials discovery pipeline. We provide a set of publicly available tests to help facilitate this effort for future model development.

Preprint – C. Bartel, A. Trewartha, Q. Wang, A. Dunn, A. Jain, G. Ceder, 2020, *arXiv* 2001.10591

Code – https://github.com/CJBartel/TestStabilityML

# 605068 Physically Informed Deep Learning for Accelerated Photosensitizer Discovery

**Abstract:**

Photodynamic therapy is arising as a non-invasive treatment modality for cancer and other diseases. One of the key factors to determine the therapeutic function is the efficiency of photosensitizers (PSs). According to the latest studies, reducing the energy gap ($\Delta E_{ST}$) between the lowest excited singlet state ($S_1$) and lowest excited triplet state ($T_1$) was an effective approach to yield highly efficient PSs and $\Delta E_{ST}$ could be fine-tuned through precise molecular donor-acceptor engineering. There is a growing literature where small families of small-$\Delta E_{ST}$ PS compounds are explored experimentally, or purely computationally at a slightly larger scale. Here, we report a large-scale data-driven search for novel PSs, with a special focus on red PS. This work employed a collaborative approach encompassing computational quantum chemistry, machine learning, organic synthesis, and nanoagent fabrication and testing.

By providing the list of commonly used donors, acceptors and bridges, there is a chemical design space of more than 7 million possible PSs. This is a vast search space that cannot be explored by chemical intuitions or Edison's approaches. Even performing quantum calculations of this vast space can take infeasibly long time. A data-driven approach by screening the $\Delta E_{ST}$ of 7 million PSs with a well-trained deep learning model provides a possible solution. Herein, we have created a high-throughput quantum calculation toolset for the construction of a novel dataset of 10k PSs with $\Delta E_{ST}$ information. This training dataset contains representative donors, acceptors and bridges with both Donor-Acceptor and Donor-Acceptor-Donor structures. A molecular graph convolutional neural network with molecular structure as features is trained on this dataset. In combination with a high-throughput PS generator, we could efficiently predict the $\Delta E_{ST}$ of possible PSs within the chemically intuited design framework. In addition to this physically informed design model, we have utilized a junction tree variational autoencoder as another deep learning method for the direct generation of possible PS candidates. These two approaches will be compared with results discussions. Furthermore, the experimental validation of chosen PSs with low $\Delta E_{ST}$ is conducted and presented. Finally the potentials and limitations of this work will be discussed.

# 610293 Quantum Chemistry-Informed Active Learning to Accelerate the Design and Discovery of Sustainable Energy Storage Materials

**Abstract:**

We employed Density Functional Theory (DFT) to compute oxidation potentials of 1,400 homobenzylic ether molecules to search for the ideal redoxmer design for non-aqueous flow batteries. The generated data were used to construct an active learning model based on Bayesian optimization that targets candidates with desired oxidation potentials utilizing only a minimal number of DFT calculations. The active learning model demonstrated not only significant efficiency improvement over the random selection approach but also robust capability in identifying desired candidates in an untested set of 112,000 homobenzylic ether molecules. Our findings highlight the efficacy of quantum chemistry-informed active learning to accelerate the discovery of materials with targeted properties from a vast chemical space.

# 598630 Color As a Source of Information in Liquid Crystal Sensors

**Abstract:**

Understanding the physical alignment of liquid crystals (LCs) near an interface is crucial for the design of electro-optical devices (e.g., liquid crystal displays [1]), nanoscale materials (e.g., semiconducting nanorods [2, 3]), and sensors [4, 5]. In this work, we are specifically interested in LC-solid interfaces, which provide a flexible platform for the development of sensors to detect air contaminants [6]. The working principle of such sensors relies on the alignment of LC molecules at interfaces, which can be controlled via tailored surface chemistries (e.g., metals and metal oxides) [6]. LC alignment and the evolution of such alignment can be visualized using orthoscopic polarized light microscopy [7]. This technique probes an LC sample with a beam of near-parallel rays of polarized light; LC alignment at the interface generates optical fields with different features (e.g., color and spatial textures) [4, 6, 8]. It has been recently shown that the features of these optical fields can be decoded by machine learning techniques to detect different types of air contaminants and to measure their concentration [13]. These studies also found that color plays a key role in distinguishing contaminant type and concentration.

In this work, we investigate color analysis techniques to transform macroscopic optical fields to physical parameters of interest such as the zenithal (tilt) and azimuthal (twist) angles of the LCs at the interface. This information is used to obtain insights into the physical origins of information in color fields. Alignment studies typically use a Berek compensator, which can quantify the optical retardance caused by birefringence (the difference between direction-dependent refractive indices) between the top and bottom interfaces of the sensor. Miller et. al. showed that LC alignment at the interface can be approximated using a Berek compensator [7]. However, a Berek compensator cannot measure optical retardance in a scalable manner. Specifically, it can not measure the evolution of the optical retardance for multiple samples simultaneously. A Michel-Levy chart is typically used to correlate color with optical retardance [10]. Sørensen et. al. formulated an equation to construct Michel-Levy charts quantitively [11] but they did not provide a direct equation to map color to LC alignment. Moreover, standard Michel-Levy charts do not consider the effect of different light sources. For instance, an approximation of optical retardance from samples under halogen light bulbs will not be accurate based on the standard Michel-Levy Chart.

In this work, we present a mathematical framework to map color in optical fields to LC alignments at the surface (and viceversa). The framework takes into consideration various aspects of relevance in LC-based sensor design (e.g., film thickness and light sources). We use our framework to construct a generalized Michel-Levy chart that captures optical retardance under different light sources. These capabilities allow us to analyze the effect of different light sources (e.g., halogen light vs. LED) and color space representations (e.g., RGB and LAB) on machine learning tasks. Specifically, we construct a 3D convolutional neural network to predict chemical species and concentrations from a temporal evolution of LC optical fields with varied light sources and color representations. This framework provides insights on how to co-design instrumentation (light sources) and the LC interface (film thickness) to extract maximum information and provides insights on the physical origins of information in LC optical responses. We use attribution methods (e.g., saliency maps [12]) to get insights of the governing dynamics of LC alignments.

[1] Drzaic, Paul S. *Liquid crystal dispersions*. Vol. 1. World Scientific, 1995.

[2] Karanikolos, Georgios N., et al. "Templated synthesis of ZnSe nanostructures using lyotropic liquid crystals." *Nanotechnology* 16.10 (2005): 2372.

[3] Yang, Chung-Sung, David D. Awschalom, and Galen D. Stucky. "Growth of CdS nanorods in nonionic amphiphilic triblock copolymer systems." *Chemistry of Materials* 14.3 (2002): 1277-1284.

[4] Bai, Yiqun, and Nicholas L. Abbott. "Recent advances in colloidal and interfacial phenomena involving liquid crystals." *Langmuir* 27.10 (2011): 5719-5738.

[5] Lowe, Aaron M., and Nicholas L. Abbott. "Liquid crystalline materials for biological applications." *Chemistry of Materials* 24.5 (2012): 746-758.

[6] Yu, Huaizhe, et al. "Amplification of Elementary Surface Reaction Steps on Transition Metal Surfaces using Liquid Crystals: Dissociative Adsorption and Dehydrogenation." *Journal of the American Chemical Society* 141.40 (2019): 16003-16013.

[7] Miller, Daniel S., et al. "Introduction to optical methods for characterizing liquid crystals at interfaces." *Langmuir* 29.10 (2013): 3154-3169.

[8] Lockwood, Nathan A., and Nicholas L. Abbott. "Self-assembly of surfactants and phospholipids at interfaces between aqueous phases and thermotropic liquid crystals." *Current Opinion in Colloid & Interface Science* 10.3-4 (2005): 111-120.

[9] Collings, Peter J. *Liquid crystals: nature's delicate phase of matter*. Princeton University Press, 2002.

[10] Spring, K. R., M. J. Parry-Hill, and M. W. Davidson. "Michel-Levy Birefringence Chart." *Olympus Microscopy Resource Center* (2010).

[11] Sørensen, Bjørn Eske. "A revised Michel-Lévy interference colour chart based on first-principles calculations." *European Journal of Mineralogy* 25.1 (2013): 5-10.

[12] Simonyan, Karen, Andrea Vedaldi, and Andrew Zisserman. "Deep inside convolutional networks: Visualising image classification models and saliency maps." *arXiv preprint arXiv:1312.6034* (2013).

[13] Cao, Y., Yu, H., Abbott, N. L., & Zavala, V. M. (2018). Machine Learning Algorithms for Liquid Crystal-Based Sensors. *ACS sensors*, *3*(11), 2237-2245.

# 600728 A Deep-Learning Potential for Crystalline and Amorphous Li-Si Alloys

**Abstract:**

This work investigates the ability of the deep-learning potential (DP) to describe structural, dynamic and energetic properties of crystalline and amorphous Li-Si alloys. Li-Si systems play an important role in the development of high energy lithium ion batteries. One challenge in simulating Li-Si systems is to balance the proper description of complex Li-Si interactions and the system size. Machine-learning potential paves an avenue to achieve this balance by describing complex interactions using machine-learning models. The machine-learning models enable us to investigate complex systems beyond the capability of the classical force fields using molecular dynamics simulations. We develop a DP for Li-Si systems with Li/Si ranging from 0 to 4.2 based on a vast dataset generated using the quantum mechanical calculations in an active learning procedure. Then we investigate the structural and dynamic properties of several crystalline and amorphous Li-Si systems using this developed DP. The DP can predict bulk densities, the radial distribution functions and diffusivity of Li in amorphous Li-Si systems at a level similar to the quantum mechanical calculations, while the molecular dynamics simulations are 20 times faster than the quantum mechanical calculations. We also discuss several issues when developing DP.

# 607895 Accelerating *Ab Intio*-Based Free Energy Sampling with Machine Learning

**Abstract:**

*Ab initio* Molecular Dynamics (AIMD) simulations are powerful for studying quantum mechanical phenomena, including chemical reactions, diffusion, and defect generations in materials, which require breaking and formation of bonds. Understanding not only the kinetics but also the thermodynamic properties of materials often require combining molecular dynamics with enhanced sampling techniques to capture the free energy landscape of the underlying phenomena. Unfortunately, free energy sampling with AIMD has been infeasible for large complex systems due to the computational cost of electronic structure calculations that are needed for each point in the configurational space. In this talk, we propose two approaches to overcome such bottleneck by using machine learning, namely the artificial neural network: (1) active learning of free energy surface during the adaptive biasing simulations and (2) on-the-fly learning of both *ab initio* force field and free energy surfaces during biased AIMD simulations. We demonstrate these two methods by applying them to catalytic reactions and defect migrations in covalently bonded solids.

# 599372 Shifting Paradigms: Using Molecular Dynamics with Machine Learning for the Calculation of Thermodynamic Properties of Fluids


**Abstract:**

The accurate prediction of fluid properties holds an essential role for chemical engineering, with ubiquitous applications in many areas from process design, reaction engineering to fluid dynamics. For this purpose, the current chemical engineering folklore employs empirical and/or theory-based correlations based on a very reduced number of experimental observations. The limitations are obvious, particularly when extrapolating to unknown compounds and/or thermodynamic conditions remote from those represented by the experiments. We propose a new workflow, where we strive to combine molecular simulations (validated against experiments) to deliver an extensive and reasonably accurate database of thermodynamic properties with machine learning (ML) algorithms to provide for the synthesis and correlation of the information. We suggest that this new paradigm can, in some instances, take over the classical experiment/equation of state tools currently in use in the field. As a proof-of-concept we explore the use of the SAFT- Mie force field[1]to generate critical properties, vapor pressures and saturated densities of a representative family of common fluids and correlate them with ML techniques.

In this work, we employ machine learning algorithms such as artificial neural networks (ANN) and Gaussian process regression (GPR). The pros and cons of the models created by each technique are analysed, considering the amount of data required, the accuracy of the model and other further applications and drawbacks of each technique. With prior knowledge of the underlying theory relating these thermodynamic properties, hybrid models are created in the sense of using suitable data transformations with theoretical basis.

Using the SAFT- Mie coarse-grained force field, molecules can be described as chains of identical spherical segments employing four key molecular descriptors . A key aspect of the model is that there is an existing equation of state (EoS)[2]that can reproduce the outputs of the molecular simulations with quantitative accuracy. By generating pseudo-experimental data for a range of these parameters, as a proof-of-concept, we attempt to replicate the SAFT-VR Mie equation of state with machine learning, to predict critical properties , vapour-liquid phase equilibrium properties and supercritical densities from these molecular descriptors and state properties.

The comparison of PvT properties between the EoS and the ML model provides a benchmark to the viability of different machine learned models in predicting fluid phase properties. Different properties present different complexity of machine-learned model, with critical properties taking zero state properties as input being the simplest model, while supercritical properties requiring two state properties (temperature and pressure). While both ANN and GPR models are able to correlate with high degree of statistical performance with $R^2$values above 0.99, the selection of activation functions and data transformation methods is essential. For a vapour pressure machine-learned model, the direct correlation between and results in a model that appears to give good $R^2$score, but corresponds to extremely large average absolute deviation (AAD), as pressure varies across different orders of magnitude. Alternatively, although the use of the Clausius-Clapeyron relationship and training a model to correlate and might appear worse near the critical point, the overall performance in AAD improves significantly. Similarly, selecting the correct activation function improves the curves of the

VLE envelope. Looking at individual saturated density curves, models with less smooth activation functions (such as the Matern kernel for GPR, or the ELU function for ANN) presents irregular shapes despite giving a statistically acceptable performance. This is because the machine is "informed" of the general shape of the model through the activation functions, which is essential to produce a good correlation. We show how the ideal activation function in this case for ANN would be the *tanh*activation function, while similarly the radial basis function is the ideal kernel for GPR.

A pathway towards the correlation of fluid properties from molecular dynamics simulated data to a machine-learned model is being drawn out, with applications from both machine learning techniques. Comparing the two machine learning algorithms used, GPR requires much less data to achieve quantitative accuracy, and it is a more sophisticated technique as it calculates the variance of the predicted model which is useful for error analysis. However, it becomes increasingly expensive, from a computational point of view, as the complexity of the problem escalates with more input dimensions. In this sense, ANN is a more robust and flexible technique but requires a much larger set of data to be generated. Discussion along these lines is included in the presentation.

References:

[1] Müller, E. A.; Jackson, G. (2014). Force-field parameters from the SAFT-equation of state for use in coarse-grained molecular simulations. *Annu. Rev. Chem. Biomolec. Eng.*, 5(1), 405–427.

[2] Lafitte, T., Apostolakou, A., Avendaño, C., Galindo, A., Adjiman, C.S., Müller, E.A., Jackson, G. (2013). Accurate statistical associating fluid theory for chain molecules formed from Mie segments. *J Chem. Phys.*, 139(15), 154504.

# 599712 Combining Molecular Dynamics and Machine Learning to Predict Interfacial Properties

**Abstract:**

Molecular dynamics (MD) simulations are a powerful tool for analyzing the properties and behavior of materials in solvent environments by explicitly representing atomistic degrees of freedom and accounting for intermolecular interactions. However, MD simulations can require lengthy runtimes that limit the high-throughput screening of system properties. Moreover, the analysis of MD simulation output typically requires the implementation of techniques that rely on human insight to relate the properties of a statistical ensemble to relevant macroscopic properties. Alternatively, machine learning has emerged as a valuable tool to interpret molecular simulation data. Here, we present computational techniques that use machine learning to analyze molecular simulations of complex solvent environments. We focus on quantifying the hydrophobicity of functionalized interfaces by utilizing deep learning to analyze MD-derived interfacial water configurations. By selecting an appropriate representation of the MD output data, we demonstrate that a 2-dimensional convolutional neural network (2D CNN) can infer spatial correlations between solvent molecules and map these features to thermodynamic observables. We systematically vary the representation of the MD input data to determine the impact on the accuracy of the ML predictions. Once trained, we find that the 2D CNN can predict dewetting free energies with minimal simulation input, facilitating the rapid screening of interfacial hydrophobicity for a range of surface compositions. We further compare the accuracy of alternative ML architectures, including 3D CNNs that incorporate temporal correlations in our deep learning framework and graph neural networks that incorporate the network connectivity of water hydrogen bonds. Together, these methodologies demonstrate that ML analysis of spatial and temporal correlations that are encoded within the solvent environment can rapidly predict interfacial hydrophobicity to guide materials design.

# 609813 Data Science Based Electrolyte Design for Organic Redox Flow Batteries

**Abstract:**

Ability to design materials for targeted functionality rather than traditional intuitive and trial-by-error based methods can bring enormous societal and technological progress. In this context, exploring the large space of potential materials for energy storage device is computationally intractable. In particular, the discovery of organic redox active species - which dictates the energy density, specific capacity and cycle life is critical step in realizing viable redox flow battery (RFB) technology - will be taunting challenge considering billions of organic structural possibilities. Traditional methodology relies on combinatorial approach of additives, functional groups and solvent composition to achieve higher solubility and stability of the specific redox molecule. However, considering the wide range of materials selection, it is imperative to build an artificial intelligence-based design formulation that will help us bottom up design of RFB electrolytes. In recent years, our team has been building database and scientific framework for electrolyte design formulations based on chemistry informed (e.g. solvate structure and inherent properties) machine learning approach. This framework helps us in solvent and functional group selection process, and ultimately enhance the solubility and stability through inverse molecular design. We will discuss our efforts and achievements related to the data science-based framework of electrolyte design formulations applicable for organic molecule based redox flow battery systems.
Ability to design materials for targeted functionality rather than traditional intuitive and trial-by-error based methods can bring enormous societal and technological progress. In this context, exploring the large space of potential materials for energy storage device is computationally intractable. In particular, the discovery of organic redox active species - which dictates the energy density, specific capacity and cycle life is critical step in realizing viable redox flow battery (RFB) technology - will be taunting challenge considering billions of organic structural possibilities. Traditional methodology relies on combinatorial approach of additives, functional groups and solvent composition to achieve higher solubility and stability of the specific redox molecule. However, considering the wide range of materials selection, it is imperative to build an artificial intelligence-based design formulation that will help us bottom up design of RFB electrolytes. In recent years, our team has been building database and scientific framework for electrolyte design formulations based on chemistry informed (e.g. solvate structure and inherent properties) machine learning approach. This framework helps us in solvent and functional group selection process, and ultimately enhance the solubility and stability through inverse molecular design. We will discuss our efforts and recent results related to the data science-based framework of electrolyte design formulations applicable for organic molecule based redox flow battery systems.

# 610519 Development of an Artificial Neural Network Model to Predict Electrical Conductivity of Pure Ionic Liquids and Ionic Liquid-Ionic Liquid Mixtures

**Abstract:**

Ionic liquids are room temperature ionic solvents characterized by low volatility, designer tunability, high thermal and electrochemical stability. These properties make ionic liquids ideal candidates for the replacement of organic solvents as electrolytes in Li-ion batteries. Despite their favorable attributes, high viscosity of many ionic liquids is a hindrance to exploiting their full potential for battery applications. One way to tackle this challenge is to develop novel ionic liquids by altering the identity of the cation, anion, or substituents on the ions. The enormous chemical space for such a manipulation is extremely difficult, if not impossible, to navigate efficiently via experimental or molecular simulation-based approaches. In such instances, machine learning techniques offer an exciting alternative, if sufficient data is available to train a model.

In this presentation, we will describe our efforts at developing a machine learning model based on the ionic conductivity data obtained from the NIST ILThermo Database. A feed forward artificial neural network (FFAAN) comprised of an input layer, a hidden layer and an output layer is obtained from the data on ~100 ionic liquids. We will demonstrate that the FFAAN is able to capture the data with a high degree of accuracy over five orders of magnitude. The machine learning model is then used to predict the ionic conductivity data for ~1200 pure ionic liquids obtained by combining the unique cations and anions in the database. We will also show how the machine learning model can be employed to predict the ionic conductivity for ~840,000 ionic liquids. The validity of such predictions will be demonstrated by comparing the ionic conductivity of those binary ionic liquids for which experimental data is available. Furthermore, we will report on 5000 potential binary ionic liquid-ionic liquid mixtures for which there exists a possibility for obtaining ionic conductivity higher than those of the pure ionic liquids making up the binary ionic liquid mixtures.

# 598191 Incorporating Long-Range Interactions in Machine Learning Models of Water and Aqueous Electrolyte Solutions

**Abstract:**

In recent years, atomistic machine learning models have become increasingly popular in molecular simulations, given their ability to combine the accuracy of quantum mechanical representations with the speed and efficiency of classical potentials. These models are capable of learning highly complex and multi-dimensional interactions within a local environment but face challenges in capturing long-range behavior. In this work, we train deep neural networks to interatomic Potential Energy Surfaces to construct many-body classical potentials that accurately represent the structure and dynamics of water and electrolyte systems. We introduce a formalism to represent long-range electrostatics by decomposition of the Coulombic Ewald energy from the local environment representation. We demonstrate the effectiveness of this approach by predicting electrolyte solution properties and vapor-liquid coexistence behavior, which highlights the versatility of our models in representing heterogeneous and charged systems.

# 600934 Fully Convolutional Neural Network Models for Materials Science Applications

**Abstract:**

In silico design can be an efficient and cost effective method for discovering new materials. Traditionally, quantum mechanics, classical approaches, and quantitative structure property relationships (QSAR) are used to design materials using the structure of the materials as an input. Although these methods are helpful in designing materials, they all suffer from a common drawback: they estimate properties from an input structure which usually necessitates iteration between synthesis and modeling to achieve the desired performance metric. Moreover, that process is also dependent on the ingenuity of design team to identify potential materials for computational evaluation which may be very different structurally from the benchmark material. Here we describe a generative deep learning approach to address some of these drawbacks: a fully convolutional neural network (FCNN) molecular generative model. The efficiency and reconstruction accuracy of this method was found to be comparable with Junction Tree variational autoencoder (JT-VAE) and superior to a SMILES-based variational autoencoder (CVAE). Furthermore, this approach (p-FCNN) is tuned to generate molecules which incorporate a larger fraction of heteroatoms (heavy atoms other than C's) compared to those from CVAE. Detailed analysis will be provided to understand the relative distribution of the newly generated molecules vs. the training set. In addition, the latent space generated from FC-NN will be organized according to ALogP and the response surface will be smoothed using a Gaussian process model. Gradient search will be performed on the organized latent space to extract molecules near the optimal value of ALogP. In conclusion, FCNN and p-FCNN can be considered as an efficient generative deep learning.approach to discover materials with an optimized performance metric.

# 599439 Deep Learning for Characterizing and Modeling the Dynamics of Complex Colloidal Self-Assembly Systems

**Abstract:**

Self-assembly (SA) is the process by which discrete components (e.g., nanoparticles in solution) spontaneously organize into an ordered structure [1]. The spontaneous self-organization central to SA enables "bottom-up" materials synthesis, which allows for manufacturing advanced, highly ordered crystalline structures in an inherently parallelizable and cost-effective manner. Thus, SA can create new avenues for highly scalable, economical manufacturing of novel metamaterials with unique optical, electrical, or mechanical properties [2-4].

SA is an inherently stochastic (i.e., random) process prone to kinetic arrest [2]. This leads to variability in materials manufacturing and possibly high defect rates, which can severely compromise the viability of using SA to reproducibly manufacture advanced materials. However, creating a systematic methodology to characterize and understand the highly nonlinear and stochastic SA dynamics responsible for kinetic arrest remains an open grand challenge. Previous attempts to characterize kinetically arrested states have either relied on physics-based order parameters that cannot accurately resolve structures in the face of thermal fluctuations or unsupervised learning techniques based on diffusion maps that provide incomplete physical interpretations of the underlying kinetics [5-7]. Moreover, all attempts to model SA dynamics either involve physics-based strategies based on the chemical master equation or parameter estimation strategies based on Bayesian Inference, neither of which are computationally tractable for practically sized SA systems [7-8]. Due to their strong adaptability and generalization properties in learning, strong noise tolerance, and ability to solve exceptionally large problems, deep neural networks (DNNs) are ideal candidates for investigating the complex, stochastic, nonlinear dynamics of SA systems [9].

In this work, we develop a type of DNN called a variational autoencoder [10] to translate the positions of individual SA particles into an easily interpretable low-dimensional space. The autoencoder provides an explicit mapping between the particle positions and the low-dimensional space. We extrapolate the knowledge from this mapping to understand the relevant symmetries that distinguish the different environments possibly present in the system and determine a rigorous physical interpretation of the low-dimensional space. We calculate low-dimensional free energy landscapes from SA trajectory data and use the relative free energies to identify the kinetically arrested states into which the system may assemble. We leverage a second DNN to model the evolution of the SA over time and predict the likelihood that SA systems avoid kinetic traps or reach target states under certain external conditions (e.g., temperature). The main contribution of this work is the development of a systematic framework to not only fully characterize kinetic traps, but also analyze exactly how these kinetic traps can be avoided in order to reproducibly reach high-value structures.

We demonstrate the efficacy of the proposed framework on an *in-silico* three-dimensional system of self-assembling DNA functionalized colloids that have shown enormous promise for sensing and photonics applications [11,12]. The system is especially prone to kinetic arrest due to the complexity of its underlying (and often competing) free energy surface, which includes repulsive interactions among the underlying silica particles, repulsive interactions due to ssDNA chain overlap, and attractive interactions due to ssDNA hybridization. We characterize and analyze the relative importance of these kinetic traps and demonstrate how changes in temperature (which turn ssDNA bonds "on" or "off") can be used to avoid kinetic traps and reach target structures.

**References**

[1] Whitesides, G. M., & Grzybowski, B. (2002). Self-assembly at all scales. *Science*, *295*(5564), 2418-2421.

[2] Paulson, J. A., Mesbah, A., Zhu, X., Molaro, M. C., & Braatz, R. D. (2015). Control of self-assembly in micro-and nano-scale systems. *Journal of Process Control*, *27*, 38-49.

[3] Grzelczak, M., Vermant, J., Furst, E. M., & Liz-Marzán, L. M. (2010). Directed self-assembly of nanoparticles. *ACS nano*, *4*(7), 3591-3605.

[4] Liddle, J. A., & Gallatin, G. M. (2016). Nanomanufacturing: a perspective. *ACS nano*, *10*(3), 2995-3014.

[5] Long, A. W., & Ferguson, A. L. (2019). Landmark diffusion maps (L-dMaps): Accelerated manifold learning out-of-sample extension. *Applied and Computational Harmonic Analysis*, *47*(1), 190-211.

[6] Reinhart, W. F., & Panagiotopoulos, A. Z. (2018). Automated crystal characterization with a fast neighborhood graph analysis method. *Soft matter*, *14*(29), 6083-6089.

[7] Tang, X., Rupp, B., Yang, Y., Edwards, T. D., Grover, M. A., & Bevan, M. A. (2016). Optimal feedback-controlled assembly of perfect crystals. *ACS nano*, *10*(7), 6791-6798.

[8] Lakerveld, R., Stephanopoulos, G., & Barton, P. I. (2012). A master-equation approach to simulate kinetic traps during directed self-assembly. *The Journal of chemical physics*, *136*(18), 184109.

[9] Jain, A. K., Mao, J., & Mohiuddin, K. M. (1996). Artificial neural networks: A tutorial. *Computer*, *29*(3), 31-44.

[10] Ohno, H. (2019). Auto-encoder-based generative models for data augmentation on regression problems. *Soft Computing*, 1-11.

[11] Pretti, E., Zerze, H., Song, M., Ding, Y., Mahynski, N. A., Hatch, H. W., ... & Mittal, J. (2018). Assembly of three-dimensional binary superlattices from multi-flavored particles. *Soft matter*, *14*(30), 6303-6312.

[12] Pretti, E., Mao, R., & Mittal, J. (2019). Modelling and simulation of DNA-mediated self-assembly for superlattice design. *Molecular Simulation*, *45*(14-15), 1203-1210.

# 606830 AI Based Analysis for Graphene Synthesis - from Size Measurement to SEM Image Generation

**Abstract:**

Graphene is a single atomic layer compound made up of pi (π) bonded carbon atoms. It is an excellent material used in the production of electronic and photonic components such as flexible transparent electrodes and field-effect transistors, due to its high mechanical strength, electrical conductivity, and mobility [1]. In this study, to synthesize high-quality graphene, a chemical vapor deposition experiment was conducted by using $CH_4$ as a precursor on Cu substrate at high temperatures [2]. The size of graphene, coverage, domain density and aspect ratio, which vary depending on the synthesis conditions including temperature, annealing time, growth time, and hydrogen supply were measured and analyzed. Owing to the variations in the shapes of graphene grains, a manual measuring technique is often used to calculate the size. Not only is this method laborious but also inefficient and resource intensive. As such, a faster, efficient, inexpensive, and standard measuring method was developed by employing Region-proposal Convolutional Neural Network(R-CNN): an objective detection algorithm [3]. In this strategy, the graphene grains were assumed to be hexagonally shaped. Consequently, the aspect ratio was also estimated by the model. Monochrome processed graphene imprinted SEM images were used as inputs by k-means clustering for the R-CNN training [4]. The R-CNN predicted size and aspect ratio were in the error margin of about 0-10% when validated against experimental data. Subsequently, to ascertain the relationship between the process variables and the size, coverage, and domain density, an Artificial Neural Network (ANN) algorithm was modelled [5]. A Support Vector Machine (SVM) was further developed to investigate the effect of the process conditions on graphene shapes using the aspect ratio as target variable [6]. Finally, by developing a Generative Adversarial Network (GAN) [7], hexagonally shaped graphene grains were generated as black and white monochrome images with the same domain density as in the experimental data. A complete network of all the developed algorithms (R-CNN, ANN, SVM, GAN) was created to generate the graphene images. To obtain the same image color and outlook just as in the experimental SEM images, a Pix2Pix model was developed to transform the monochrome images into colored real SEM images [8]. By comparison, the ANN predicted images showed high agreement with the experimentally generated ones. As a result, we were able to find optimum experimental condition efficiently using this newly developed model, having a deeper insight into graphene CVD synthesis and predicting results in the form of images.

Literature cited:

[1] Phaedon Avouris and Fengniam Xia, "Graphene applications in electronics and photonics" ,MRS Bulletin,vol.37,pp. 1225-1234, 2012.

[2] Xuesong Li, Wei Wei Cai, Jinho An, Dongxing Yang,Richard Piner, Aruna Veamakanni, Inhwa Jung, Rodney S.Ruoff "Large-Area Synthesis of High Quality and Uniform Graphene Films on Copper Foils", *Science*, vol. 324, pp. 1312-1314, 2009.

[3] Ross Girshick, Jeff Donahue, Trevor Darrel, Jitendra Malik "Rich Feature Hierarchies for Accurate Object Detection and Semantic Segmentation", The IEEE Conference on Computer Vision and Pattern Recognition (CVPR), pp. 580-587, 2014

[4] Ray Siddheswar, Turi R.H. "Determination of number of clusters in k-means clustering and

application in colour image segmentation", Proceedings of the 4th international conference on advances in pattern recognition and digital techniques, pp.137-143, 1999

[5] Kenji Suzuki, "Artificial Neural Networks - Methodological Advances and Biomedical Applications", ISBN 978-953-307-243-2, pp. 2-14, 2011

[6] Hsu, C.-W., Chang, C.-C., and Lin, C.-J. "A practical guide to support vector classification. Tech. rep., Department of Computer Science", National Taiwan University, 2003

[7] Goodfellow, Ian J, Pouget-Abadie "Generative adversarial nets", Advances in Neural InformatiProcessing Systems, pp.2672-2680, 2014

[8] P. sola, Phillipzhu, Jun Yan Zhou "mage-to-Image Translation with Conditional Adversarial Networks", Proceedings - 30th IEEE Conference on Computer Vision and Pattern Recognition, pp. 5967-5976, 2017

# 602682 Infrared Spectra Prediction with Machine Learning

**Abstract:**

Spectroscopy is an important tool for manufacturing for ascertaining purity in biological systems and drug manufacturing. Prediction of unknown spectra would be a powerful tool for extending existing spectral identification databases and evaluating novel molecules for which there exist no standards. Machine learning algorithms using deep neural networks have been used for the interpretation of spectra in microbiology and food production for the last ten years. The prediction of the spectra themselves is a more recent area of research, with new efforts to predict IR spectra, proton NMR spectra, and mass spectra. It is the purpose of this present effort to use machine learning to predict IR spectra for novel molecules, improving on previous such work by including effects beyond calculated harmonic frequencies and using learned molecular fingerprints.

A model for IR spectral prediction has been developed, requiring only the input of a molecular graph structure provided in the form of a SMILES code. The chemical representation is processed through a message passing neural network followed by a feed forward neural network for prediction of IR spectra, an adaptation of the Chemprop code for molecular property prediction. The model is pre-trained using quantum chemistry calculations for molecules sampled from the PubChem database to learn the molecular harmonic vibration modes. During pretraining, active learning is used to explore the chemical space more efficiently and prioritize molecules which will most improve the model. Further training is performed using experimentally collected spectra available in open-access databases from the National Institute of Standards and Technology (NIST) and the National Institute of Advanced Industrial Science and Technology (AIST). The model allows for predictions of spectra in the gas phase and in supported condensed phases.

# 607833 Optimization of Thermal Conductivity at Interfaces Using Learning Algorithms

**Abstract:**

In material science, we are frequently interested in understanding the properties and design implications of material at interfaces. These interfaces can be manipulated to improve the desired characteristics of the bulk material. Optimization of thermal transport across the interface of two different materials is critical to micro/nanoscale electronic, photonic, and phononic devices. However, interfacial thermal conductivity optimization is a high-dimensional problem due to the number of factors affecting it. Non-equilibrium molecular dynamics (NEMD) has been used to investigate the thermal transport across a solid-solid interface between different materials. Despite NEMD being an effective approach to this problem, scanning the large parameter space is unrealistic due to high computational cost. Alternatively, machine learning is a cost-effective and time-efficient method that can investigate this interfacial thermal transport phenomenon.

In this talk, we describe our efforts to understand and optimize the impact of interfacial atomic defects and intermixing on the thermal transport across an Al/Si junction. We developed a learning-based framework to optimize over a parameter space that would be prohibitively large in the brute force approach. Using this technique allows us to accumulate knowledge of the system of a given type of atoms and store this information into a neural network. A scalable framework was built to allow utilization of the available high-performance computing resources, thereby accelerating the learning by directly interfacing with the LAMMPS molecular dynamics simulation package. We found that mixing 3-4 monolayer of atoms near the interface can significantly increase the thermal conductance. However, when inter atomic mixing layers get thicker, it induces a negative effect on the thermal conductance due to phonon scattering. The model is able to find the optimal mixing length and mixing fraction relatively fast. This work can be extended to more systems by inclusion of descriptors and a high-fidelity surrogate model.

# 595965 High-Throughput Approach to Studying Adsorption on Zeolite Surfaces

**Abstract:**

Zeolites are nowadays abundantly used as adsorbents and catalysts for many industrial applications. Computational studies have recently complemented the widely available experimental data and observations on zeolites by shedding light on reaction mechanisms and the active sites. The computational studies of zeolites, however, are challenging due to the numerous possible active sites. A zeolite structure can have many unique tetrahedral sites (T). Substitution of a trivalent aluminum atom to a tetravalent silicon atom in the zeolite framework leads to a net negative charge that can be compensated by a proton or a cation. However, there are numerous possibilities for the placement of Al and the compensating proton or cation. The number of possibilities grow exponentially for higher Al/Si ratios in the framework. Due to this complexity, most theoretical studies have focused on a limited number of T sites based on physical intuitions. For example, many studies of the MFI zeolite, which has 12 T-sites, have focused only on one or two T sites based on the accessibility of guest molecule to be the active site. These choices are justifiable given the high computational cost of accessing the full descriptor space, however, they probably miss promising active sites. In this work, we attempt to thoroughly explore the possible active sites using a high-throughput approach. We apply our approach to the problem of NO adsorption on metal-exchanged zeolites. It is known that Pd-exchange CHA, for example, is an effective adsorbent to limit NOx emissions. Full exploration of this problem requires enumerating the possible Al-Al pairs on the starting zeolites, identifying the favorable sites for the compensating protons and cations and searching for a global minimum in each case, which means thousands of calculations even for simple zeolites. The problem is further compounded when the cation is a metal that might have multiple oxidation states. We show how we systematically explore the possible sites, starting with a lower level of theory on the initial calculations, and then narrowing the potential candidates and using a higher level of theory afterwards. The automation of this calculation reduces the amount of manual labor and highly improve the efficiency of the workflow. It presents a new approach to tackle the complicated problems related to zeolite catalysis.

# 607612 Machine Learning-Guided Flow Synthesis of Inorganic Metal Halide Perovskite Quantum Dots

**Abstract:**

The controlled synthesis of advanced colloidal nanomaterials is limited by the currently available reactors utilized for their synthesis and the disjointed workflow between experimentation and data processing. Manual flask-based experimentation strategies, although applied across most research laboratories, are unlikely to sufficiently capture and control the complex and highly sensitive reaction systems of many colloidal nanomaterial syntheses. These batch reactors are slow, labor-intensive, and subject to batch-to-batch variations in experimental data sets. In response, over the last decade different microfluidic synthesis strategies have been developed and effectively applied towards the controlled synthesis of different classes of colloidal nanomaterials. Such flow synthesis strategies notably enhance the efficiency of colloidal synthesis studies, while simultaneously unlocking previously unattainable precision in synthesis parameter control. Despite the effectiveness of flow synthesis microreactors in accelerating reaction condition screening of colloidal nanomaterials, they still rely on user-directed experiment selection. Broad reaction space screening using design of experiments is materially inefficient, and expertly guided experiments are often too slow to effectively cover a synthesis space within the lifetime of unstable chemical precursors. These bottlenecks can be addressed through development of self-optimizing fluidic reactors and integrating fully automated experimentation, materials diagnostics, and data analysis with artificial intelligence-guided experiment-selection algorithms. Such autonomous materials synthesis platforms will be able to isolate optimal reaction parameters entirely without user intervention.

Colloidal semiconductor nanocrystals (known as quantum dots) are a particularly suitable testbed for autonomous materials development strategies. Quantum dot syntheses possess a large and highly sensitive reaction parameter space.[1] Proper control of their reaction conditions is difficult to attain both within a single set of precursors and across research labs (batch to batch variation). As a result, despite extensive efforts and constantly growing applications, many synthesis paths are either unexplored or not fully optimized. One of the recently emerging classes of quantum dots is colloidal metal halide perovskites, which have highly favorable properties for applications in optoelectronic devices. Due to the combinatorially increasing synthesis space of metal halide perovskite quantum dots, many of their possible synthesis routes are novel and unexplored. Therefore, these reaction systems present an exciting opportunity for exploration within an autonomous flow synthesis platform.

In this work, we present the first machine learning guided-fluidic reactor for the controlled synthesis of high-quality colloidal quantum dots. While the strategies and techniques detailed here may be easily applied to a large number of solution-processed materials, as a case study we analyzed the halide exchange of cesium-lead-bromide perovskite quantum dots with zinc halide salts and two different surface ligands.[2] We developed a novel *in situ* material diagnostic module to attain accurate real-time access to peak emission energy, emission linewidth, and photoluminescence quantum yield of the in-flow synthesized perovskite quantum dots. These output parameters, corresponding to the physicochemical and optoelectronic properties of quantum dots, were then simultaneously optimized through an objective function integrated within a custom neural network ensemble model and various decision-making policies (used to select conditions from the model). Utilizing the developed autonomous materials development platform, we evaluated the performance of eight different experiment selection algorithms (spanning machine learning to evolutionary algorithms) over the course of more than 1400 experiments. The controlled syntheses of perovskite quantum dots were performed without user intervention and required less than 400 μL of starting quantum dot solution per experimental condition. The methods detailed in this work offer autonomous reaction optimization across multiple vital performance parameters. Further use of similar self-optimizing systems in the field of colloidal nanoscience would reduce the cost and expedite the process of materials development and discovery.

## References


[1] R. W. Epps, K. C. Felton, C. W. Coley, M. Abolhasani, *Lab Chip* 2017, *17*, 4040.

[2] K. Abdel-Latif, R. W. Epps, C. B. Kerr, C. M. Papa, F. N. Castellano, M. Abolhasani, *Adv. Funct. Mater.* 2019, *29*, 1900712.


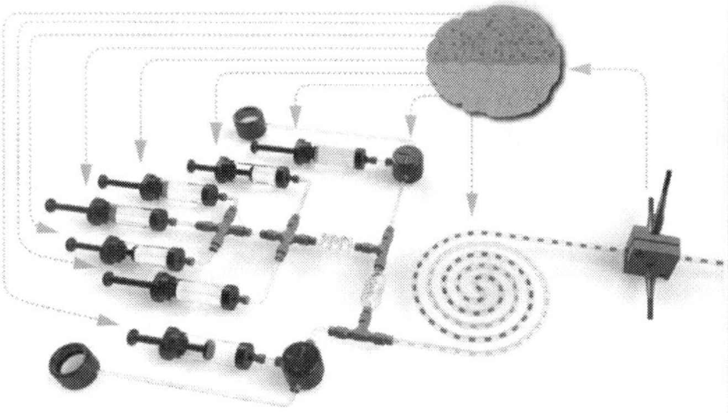

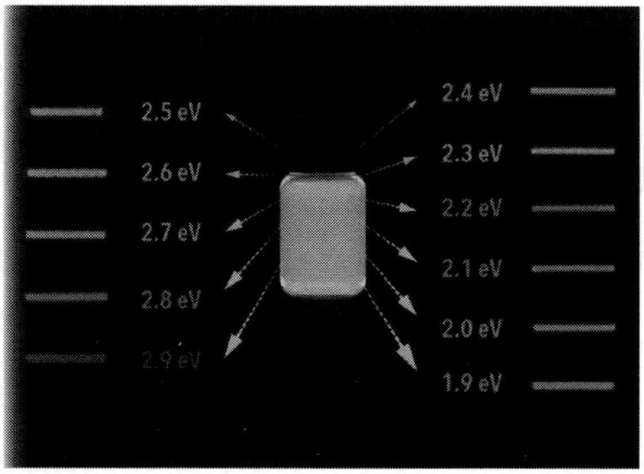

# 609449 High-Throughput and Data-Driven Strategies for the Design of Deep Eutectic Solvent Electrolytes

**Abstract:**

Within the framework of green chemistry, Deep Eutectic Solvents (DES) have been identified as promising candidates for use in many applications, including battery electrolytes. DES are characterized by two or three materials that associate with each other through hydrogen bond interactions, resulting in a eutectic mixture whose freezing point is below that of the individual materials. This design space is overwhelmingly large and poses a challenge for screening a vast and diverse set of materials. Here we present a strategic approach consisting of high throughput experimentation (HTE) coupled with data science driven analysis to identify exceptional DES candidates based on key physiochemical and electrochemical properties. Much of our HTE adopts methods that are already used frequently in the biotech and pharmaceutical industries, most notably performing parallel syntheses and analyses in 96-well-plate formats. DES samples are first synthesized using an open-sourced automated liquid handling robot. DES melting points are then determined by monitoring the melting process with an infrared camera and identifying the temperature at which the thermal conductivity of the samples changes abruptly. The solubility of battery redox-species is determined via UV-VIS well-spectrophotometers. Finally, the electrochemical stability window and cycling properties of DES electrolytes are measured in high-throughput by using screen-printed electrodes on 96-well plates adapted for use with a standard potentiostat. The ability to rapidly and efficiently collect data also creates a need for the development and use of automated processes for data analysis, which have been developed in an open-sourced format by our group. This approach to HTE also allows for the incorporation of data science techniques, such as feature extraction and machine learning, that further aid in probing a design space that is ultimately too large for experimental methods alone.

# 602248 Development of a Hybrid Model to Describe the Dynamics of a Partially Known System

**Abstract:**

Mathematical modeling plays a critical role in the design, optimization, and control of chemical processes [1]. In order to apply models to a wide range of process conditions, it is often desirable to describe the system using a first principles model. However, the formulation of such models can be challenging if the underlying mechanisms are only partially understood [2, 3, 4]. Under such circumstances, a data-driven model can be built from measured data, and the resultant model can describe the system adequately, even when mechanistic understanding is limited [5, 6]. However, a data-driven model has narrow applicability as it is tailored to describe the input-output relationships contained in the training datasets [7]. As an alternative, a hybrid modeling approach that combines first-principles and data-driven modeling has been proposed to describe a partially known process [7, 8]. Hybrid models have better prediction capabilities than first-principle models, and thus will have better generalizability and interpretability than data-driven models.

In this work, we proposed a new numerical scheme to construct a hybrid model for a system that has partial mechanistic information. A hybrid model was constructed by introducing correction terms for estimating process-model mismatches between mechanistic model predictions and experimental measurements [9]. Here, a nominal model refers to a mechanistic model based on our prior knowledge of the system of interest, and the correction terms are assigned to differential equations of the nominal model so that the resultant hybrid model has improved prediction capabilities. The proposed methodology consists of three steps. First, observability analysis was performed for the nominal model to determine a subset of states that are observable from the available measurements. This step is critical to reduce the computational load of the estimation and avoid potential overfitting issues. Second, the available measurements were used to estimate the correction terms associated with the observable states through solving a L2-regularized least-squares problem [10]. Lastly, once the correction terms were estimated, the trajectories of the states were simulated based on the hybrid model. Then, an artificial neural network (ANN) model was developed for predicting the correction terms for a given set of state values, and the developed ANN was used within the hybrid model. Through incorporating the ANN, the hybrid model was able to predict dynamic responses to an input, which was not in the measurement data used for estimating the correction terms.

The validity of the proposed method was demonstrated by constructing a hybrid model for the NFκB signaling pathway induced by both lipopolysaccharide (LPS) and brefeldin A (BFA). The signaling dynamics induced by LPS have been well studied, and hence, a previously published model was used as the nominal (mechanistic) model [11]. Since the dynamics initiated by BFA are less characterized, the nominal model could not capture the signaling dynamics induced by both BFA and LPS well. By implementing the proposed methodology, the developed hybrid model was able to predict the NFκB signaling dynamics induced by both LPS and BFA accurately.

References

[1] Hangos, K. M.; Cameron, I.T. Process modelling and model analysis. Academic Press: London, 2001.

[2] Cho, K.-H.; Choo, S.-M.; Jung, S.H.; Kim, J.-R.; Choi, H.-S.; Kim, J. Reverse engineering of gene

regulatory networks, IET Systems Biology 2007, 1, 149 - 163.

[3] Willis, M.J.; von Stosch, M. Inference of chemical reaction networks using mixed integer linear programming, Computers and Chemical Engineering 2016, 90, 31-43.

[4] Lee, D.; Jayaraman, A.; Kwon, J.S. Identification of a time-varying intracellular signalling model through data clustering and parameter selection: application to NF-κB signalling pathway induced by LPS in the presence of BFA, IET Systems Biology 2019, 13, 169-179.

[5] Thompson, M.L.; Kramer, M.A. Modeling chemical processes using prior knowledge and neural networks, AIChE Journal 1994, 40, 1328-1340

[6] Narasingam,A.; Kwon, J.S. Development of local dynamic mode decomposition with control: Application to model predictive control of hydraulic fracturing, Computers & Chemical Engineering 2017, 106, 501-511.

[7] von Stosch, M.; Oliveira, R.; Joana Peres, J.; de Azevedo, S.F. Hybrid semi-parametric modeling in process systems engineering: Past, present and future. *Computers & Chemical Engineering* **2014**, *60*, 86-101.

[8] Faizan Bangi, M.S.; Kwon, J.S. Deep hybrid modeling of chemical process: Application to hydraulic fracturing, Computers & Chemical Engineering 2020, 134, 106696.

[9] Engelhardt, B.; Fröhlich H.; Kschischo, M. Learning (from) the errors of a systems biology model, Scientific Reports 2016, 6, 20772.

[10] Howsmon, D.P.; Hahn, J. Regularization Techniques to Overcome Overparameterization of Complex Biochemical Reaction Networks, IEEE Life Sciences Letters 2016, 2, 31 – 34.

[11] Lee, D.; Ding, Y.; Jayaraman, A.; Kwon, J.S. Mathematical modeling and parameter estimation of intracellular signaling pathway: application to LPS-induced NFκB activation and TNFα production in macrophages, Processes 2018, 6, 21.

# 607279 Ffiber: Framework for Fluorescent Neuroimaging Based Experimental Routines

**Abstract:**

**Background:** The modern lab produces more data than it can analyze using traditional experimental techniques. For neuroimaging labs, terabytes of imaging data can easily be generated in the course of a single experiment. However, only a small portion of this data is accessible for extracting results based on existing methods. To enhance the portion of data that can be analyzed, many labs are turning to data science for high throughput processing and analysis. However, to synthesize highly skilled domain scientists with an effective data science strategy, a robust method for developing well-designed and data informed experimental pipelines must be followed. We developed a framework for fluorescent neuroimaging based experimental routines (FFIBER) that implements key practices of data science, software engineering, chemical engineering, and neuroimaging. The goal of this pipeline is to produce experimental workflows for high throughput neuroimage acquisition and analysis. For each experiment we follow a six-step process: (1) develop a data awareness, (2) design a data management plan, (3) determine an optimal experimental pipeline, (4) build out supporting data science infrastructure, (5) perform primary and supplemental imaging, and (6) produce interpretable visualizations of results. The *objective* of this project is to apply our FFIBER method during initial experimental stages and retroactively to generate, process, and analyze high volumes of neuroimaging data without an overall increase in labor-hours.

**Methods:** To develop a data awareness, we performed a data assessment of relevant metadata, data storage, and data relationships. When designing a data management plan, we identified the optimal data storage locations, file structure for raw data and results, and personnel responsibilities. To determine an optimal experimental pipeline, we performed software reviews, addressed labor-hour bottlenecks, and identified necessary programs, packages, and scripts. Then we built out the supporting data science infrastructure utilizing Python and Jupyter notebooks as the base script for pipelines, electronic laboratory notebook integration, and version control with the team through GitHub and Google Drive. After the pipeline was built, we completed the primary set of experimental imaging using confocal microscopy of brain slices and fed the data through the pipeline for analysis. We then completed supplemental imaging to experimentally support a representative set of results obtained through FFIBER. Finally, we wrote Python scripts to output consistent interpretable visualizations of our analyses and align these with other experimental results.

**Results:** We applied FFIBER to a data science experiment analyzing the phenotype of microglial cells in *ex vivo* brain slices to assess extent of glial cell activation. We were able to create and modify a pipeline during the initial stages of (1) an oxygen-glucose deprivation experiment and retroactively for (2) an inflammation-sensitized model using *E. coli derived* lipopolysaccharide, (3) and a neonatal hypoxia-ischemia (HI) model. A preliminary data assessment revealed a storage necessity of 5 GB per experiment with five individual file types from five separate personnel. We completed the data assessment and built a data management plan with university-supported Google Drive as our storage location, *.tiff*, *.csv,* and *.ipynb* as our file types, and established a slice-based, double-blind file structure. We wrote Python scripts automating image upload, cleaning and automatic feed to our analysis pipeline that reduced labor-hours from five hours per image set to less than ten minutes. Our software review led us to integrate a published cell morphometrics package, VAMPIRE, with our diff_register package for skeletonized morphometrics of microglia. Our image analysis pipeline developed with FFIBER split uploaded confocal images into fourths, segmented, skeletonized and identified the microglia in the images, and uploaded the images into our integrated morphometric analysis. FFIBER structured the development of our work pipeline that decreased work hours from a full day to less than an hour to analyze an entire experiment. Finally, our FFIBER structured pipeline enabled us to detect and quantify shape differences between non-treated and injured *ex vivo* slices – providing additional information for our initial oxygen-glucose deprivation experiment and retroactive insight to our inflammation sensitized and neonatal HI models. Our first application of FFIBER decreased our human labor time from > 24 hours per 5 GB image set to less than an hour, decreased our storage budget to $0, and produced information about cell morphometrics previously unavailable to our lab due to lack of analysis software and time constraints.

**Conclusions:** By applying FFIBER, we were able to create workflow pipelines both during initial stages of experimental design and retroactively in neuroimage sets already collected. We decreased labor hours for data sets from 24 hours to less than an hour through using Python scripts rather than manual work and a well-designed pipeline. Our framework creates a high throughput approach for experimental design in research dependent on neuroimaging by integrating traditional experimental methods with modern data science applications. We can further iterate the methodology to include metadatabases for easier data access and data connections, budget vs storage matrices for determining best data storage location, and templates for electronic lab notebooks, file structures, data relationships, and generalized processing pipelines. Additionally, while FFIBER was developed for fluorescent neuroimaging datasets, the current workflow is robust, and flexible enough for application to other image analysis techniques, other organs, and other fluorescent imaging methods.

# 597491 Data-Driven Discovery of Polymeric Vehicles for Gene Editing:Serendipity-Inspired Design Directions.

**Abstract:**

Synthetic materials such as cationic polymers are poised to replace engineered viruses as vehicles for therapeutic nucleic acids. Polymeric materials offer a vast design space where desired material properties can be engineered by deft optimization of chemical composition and polymer architecture. However, rational polymer design is impeded by the "curse of dimensionality" since systematic examination of numerous variables such as composition, architecture, length and formulation parameters is paramount. To accelerate polymeric vector discovery, we need to develop models that predict the impact of key design attributes such as phase behavior, basicity, surface charge, nucleic acid binding, and hydrophobicity on the ultimate biological responses observed. In this work, we eschew low-throughput hypothesis-driven design of polymers in favor of a data-driven approach that exploits combinatorial polymer libraries and high-throughput experimentation. Thereafter, we employ statistical learning to extract the molecular attributes of the hit polymers that promoted gene editing efficiency and apply this knowledge to provide experimental guidance on designing subsequent polymer libraries.

Drawing inspiration from high-throughput experimental approaches used in the pharmaceutical industry to screen drug libraries for therapeutic activity, we employed parallel polymer synthesis, formulation and well plate based biological assays for rapid screening of gene editing efficiency We synthesized a chemically diverse library of copolymers combining different ratios of 1) cationic monomers bearing amines spanning a broad range of basicity and 2) neutral monomers of varying hydrophilicity. Our combinatorially designed library allows for systematic investigation of the effect of amine basicity by studying variations in polymer pKa resulting from the use of 4 cationic monomers, in ratios of 100, 75, 50 and 25 %. Subsequent to parallel polymer synthesis, extensive physicochemical characterization was completed using automated tools -: composition and molecular weight analysis, $pK_a$, polyplex size distribution, binding assays, and $\zeta$-potential measurements were acquired in high-throughput modes to generate a rich dataset. Polymers were complexed with ribonucleoprotein (RNP) payloads and gene editing was quantified by estimating the proportion of mCherry positive cells. To resolve the trade-off between sensitivity and experimental throughput, we employed image cytometry and developed an image processing algorithm to quantify mCherry expression in a robust and automated fashion from a bank of images arising from 200 unique formulations. At the end of the high-throughput screening campaign, we obtained a high-performing hit polymer that outperformed state-of-the-art synthetic transfection reagents.

Having identified a hit formulation from our library, the challenge was to: 1) unravel the relationship between polymer attributes and gene editing efficiency. 2) build on these structure-activity relationships to guide the design of future polymer libraries that will yield a higher "hit rate". In order to derive these predictive relationships, we sought to understand how 10 polymer descriptors influenced biological performance. We turned to principal component analysis (PCA), to deal with the dimensionality challenge created by our dataset. Through PCA, we concluded that a single polymer descriptor cannot be used in isolation to guide the design of future libraries. Rather, complex non-linear relationships between several molecular attributes were responsible for editing performance. Discarding preconceived notions of how various chemical functional groups will influence delivery, we screened a large chemically diverse polymer library to discover design guidelines that do not conform to traditional heuristics as well as a promising hit polymer, which may not have been accessible through hypothesis-testing. If statistical learning and automated experimental workflows are applied in tandem, we can overcome challenges originating from the complexity of structure-function relationships governing polymeric gene delivery.

# 605093 Artificial Intelligence-Assisted Full-Map Understanding of Strain Sensing Devices and Designated Predictions for Soft Robotic Systems

**Abstract:**

Compliant strain sensing devices are widely applied in real-time mechanical feedback for soft robotic systems. However, it remains challenging to reach customized sensing devices with high sensitivity coordinated with designated working windows due to the complex and unclear structure-composition-property relationships which are hard to predict and pre-design. In this work, through active learning and developed algorithms, we propose machine assisted full-map structure-composition-property understandings of strain sensors consist of classical nanocomponents of $Ti_3C_2T_x$ MXene, carbon nanotube, and poly(vinyl alcohol). Important variables of nanocomponent weight loading, nanocoating thickness, and nanocoating surface morphology are systematically investigated, and an informative data base (5000+ experimental resistance points, 500+ nanocoatings, and 200+ sensing devices) is constructed for training of machine learning models. Besides, comprehensive algorithm optimizations are performed for better machine performance. Ultimately, the machine learning models show low average scattered numerical error of 24% and give customized predictions in 1) sensing performance in a given material recipe; 2) suitable material recipes of sensing devices with high sensitivity and designated working windows in soft robotic systems. The reported strategies here open new perspectives towards design-free high-performance strain sensing devices and customized applications in soft robotic systems.

More specifically, in our material systems, at least 33,700 non-repeating recipes are available for strain sensing devices fabrication although we limit the step length in each variables of material recipes. With such a complex system and very limited understanding, there are no available *in silico* simulation methods to interpret or predict the sensing devices. In addition, it is nearly impossible to screen the whole design space *via* Edison approach because the sensing needs are customized every time. Herein, we have proposed a data-driven framework to exploratively learn the full-map of the strain sensing devices. The data is collected with thirteen active learning rounds which investigate both the mean minimal distance ($L_2$) between suggestion points and existing data points, and the numerical uncertainty of well-trained models during each round. After a representative dataset is constructed, optimization strategies including GA combination, customized loss function, and data augmentation are used for improving the model performance. This best model has achieved a low average scattered numerical error ((Predicted value - Real value)/Real maximum value in device) of 24%.

# 608059 An Autonomous Computational Workflow for Efficient Generation of Polymer Data

**Abstract:**

Data development is critically important for the emerging polymer informatics ecosystems. Two primary challenges of preparing a computational polymers database are which polymer (out of a huge number of possible candidates) should be considered and how to obtain a reasonable polymer model, from which calculations for relevant can be performed. We have developed an autonomous computational workflow, involving an active-learning guidance step and an efficient Polymer Structure Prediction strategy, to solve these problems. Our scheme has been implemented in the Computational Autonomy for Materials Discovery package and used to significantly enlarge the computational polymer database that power the Polymer Genome platform (https://www.polymergenome.org/). Machine-learning models trained on our computational data for predicting various polymer properties are available in https://www.polymergenome.org/. This work is supported by Toyota Research Institute and Office of Naval Research.

# 597874 Leveraging Experimental Transition Metal Complex Information to Improve Generalizability of Machine Learning Models

**Abstract:**

Creating computationally inexpensive, yet accurate, models for materials and molecular properties is a primary driver for data science in chemical engineering. However, extending these models to target systems beyond the training data often requires models to aggregate large quantities of information from different chemistries. Our group has previously generated accurate artificial neural network (ANN) models trained on density functional theory (DFT) properties for mononuclear octahedral transition metal (TM) complexes primarily through exhaustive enumeration. Here, towards expanding the applicability of our ANN models to more diverse chemical space, we first mined all experimental, structurally identified mononuclear octahedral TM complexes in the Cambridge Structural Database (CSD). By featurizing these complexes with revised autocorrelation functions, a class of graph-based heuristic descriptors developed in our group, we develop a bird's eye view of their similarity to each other and to prior data generated in our group's database. Next, we compare a number of approaches for selecting the next CSD complexes to evaluate by DFT, with the aim of accelerating the generalizability of re-trained ANN models to these diverse chemistries. Approaches include physically-motivated sampling, such as most-diverse sampling of ligand symmetries, and ANN-derived uncertainty sampling, i.e., selecting complexes with the lowest model confidence. We anticipate our benchmarked methods for selecting complexes for evaluation will be useful across domains where rapid expansion to unseen materials is desired.

# 605838 A Formal Grammar-Based Machine Learning Approach for Predicting Reaction Outcomes

**Abstract:**

Prediction of reaction outcomes without performing time-consuming experiments offers several invaluable advantages -- reducing experimentation costs due to elimination of highly improbable reactions, lowering the time to hypothesis-validation-and-correction, enabling high-throughput experimentation for complex reactions, and allowing significantly more time for analyzing the results. In recent years, machine learning techniques have been successfully applied in this area that, either directly or indirectly, address one or more of the above issues. A few areas of application include discovery of novel molecules with desired properties, understanding structure-property relationships, identifying catalysts for better reaction catalysis, and data-driven modeling of chemical reaction kinetics. The success of machine learning methods could largely be attributed to their inherent proficiency in capturing complex non-linear dependencies between various factors that govern the reaction systems. Here, we present an approach for predicting the most likely product of a reaction based on a given set of reactants and agents (catalysts, reaction medium). The structural properties of reactants and the product molecules are encoded based on a formal grammar, akin to context-free grammars (CFG) in the area of natural language processing (NLP), that essentially gives rise to a parse-tree representation of the molecule. This representation is used in a convolutional neural network architecture to model the structural transformations from a set of reactants to the most likely product of the reaction. The proposed framework could be used for rapid experimentation and analysis of reaction outcomes, especially if the molecules involved are less-known and structurally novel. In addition, the latent space of the neural network architecture could also be used to discover insights on the transformations from reactants to products in a given reaction under different conditions.

# 611660 Inverse Learning of Material Physics through Image Data and Continuum Modeling

**Abstract:**

The advent of big data in spatio-temporal microscopic images as well as large-scale atomistic simulations offers a tremendous amount of hidden information about the physical chemistry at the continuum level. Using a framework of PDE-constrained optimization, we extract multiple unknown constitutive relations simultaneously from a small set of images of pattern formation. Compared to the data-driven modeling approach, our approach provides clear physical interpretability by prescribing a general governing equation while faithfully achieving quantitative matching between data and continuum model prediction through highly expressive nonlinear and/or nonlocal (integro-differential) constitutive relations.

We present examples of learning state-dependent properties such as the diffusivity, kinetic prefactor, free energy, and direct correlation function from Cahn-Hilliard equation, Allen-Cahn equation, and Phase Field Crystal Model. Applying the method to experimental images of lithium concentration field in lithium-iron phosphate (LFP) particles during charge and discharge, we are able to extract its free energy and reaction kinetics, and "image" its reaction kinetics heterogeneity map. In addition, we also demonstrate its ability to coarse-grain molecular simulations and hence provide acceleration by learning functions of free energy and diffusivity and achieving quantitative matching of the patterns of spinodal decomposition.

# 607537 Improved Training of Graph-Embedding Based Neural Network Energy Functions for Catalysis


**Abstract:**

Complex reaction networks comprising of thousands of intermediates and reactions are ubiquitous in heterogeneous catalysis; developing mechanistic models of such systems is computationally intractable primarily due to the cost of computing the energies of reactions and species using an ab initio method such as density functional theory (DFT). Developing machine learned energy potentials trained on DFT energies offer a tractable way of building mechanistic models for these systems. To this end, the contribution of this talk is two-fold. First, we develop a novel deep neural network model to learn and predict the potential energy surface (PES) for periodic systems. The novel architecture integrates graph embedding of individual atoms, atomic interactions and periodic boundary conditions with physics-aware feature engineering, residual learning, and attention mechanism. Second, we train the model with an efficiently distributed Hessian-free optimization method and deliver a prediction within chemical accuracy, which demonstrates the potential of second order methods in the joint field of deep learning and quantum chemistry. We will use examples from the field of heterogeneous catalysis, including (i) alkane conversion on molybdenum sulfide and (ii) adsorption of hydrocarbons in zeolites.

# Explainability based Fault Detection and Diagnosis using Deep Learning

Piyush Agarwal[1], Melih Tamer[2], and Hector Budman[1]

[1]Chemical Engineering Department, University of Waterloo, Waterloo, Canada, N2L3G1
[2]Sanofi Pasteur, Toronto, Ontario M2R 3T4, Canada

April, 2020

The fourth industrial revolution also known as 'Industry 4.0' and Big Data revelations has enabled the manufacturing industries to boost its performance in terms of operation, economic and safety. In order to meet high expectations of quality, efficiency and reliability of the product, a process monitoring system is needed. The two important aspects of Statistical Process Monitoring (SPM) are fault detection and diagnosis (FDD). Safe operation of process plants can be ensured by determining if the current state of the process is normal or abnormal. This problem is known as "Fault Detection". Following the detection of abnormality in the process, the next step is to classify the fault and diagnose its causes, also known as "Fault Diagnosis/ Classification". Noisy and high dimensional recorded process inputs greatly hinders the FDD mechanism in process plants. Previously, traditional linear PCA and its variants [5, 6] based methods have been extensively used for fault detection and prognosis. However, the inherent non-linearity in the process poses challenges for these methods. To this end, deep learning methods have shown considerable improvement over traditional methods [3, 4]. Deep learning methods can be broadly classified into two categories: 1) Unsupervised Learning; 2) Supervised Learning. In this work we use a supervised approach on a benchmark Tennessee Eastman Process (TEP) and show the improvement over both PCA-PLS based methods, unsupervised learning and previously used supervised learning based methods. This work is motivated by the fact that deep learning algorithms require generally large amounts of data for proper training and thus, their accuracy may be compromised when using relatively small datasets (bottleneck for many industries such as pharmaceutical industries etc.). Therefore, there is a large incentive to increase the accuracy of deep learning based FDD models without increasing the amount of data for training. Towards this goal the concept of explainability of Neural Networks (NNs) is exploited to improve the accuracy of the FDD without increasing the amount of data. In this study, the explainability of the effect of the inputs on the output classes is quantified by relevances interpreted using Layer-wise Relevance Propagation (LRP) and the irrelevant input variables' for the supervised classification problem are discarded thus enhancing both fault detection and fault diagnosis accuracies [1, 2]. This relevance based procedure greatly reduces the number of tuning parameters in the network and also provide physical insights for the classification problem at hand. Lastly, we use Dynamic Deep Supervised Autoencoder (DDSAE) NNs which improves the classification ability compared to traditionally used SAE (Stacked Autoencoder) NNs. The efficacy of the proposed method over different methods for both Fault Detection and Diagnosis are shown in Table 1 and Figure 1 respectively.

The proposed deep learning based detection method is compared with linear Principal Component Analysis (PCA), Dynamic Principal Component Analysis (DPCA) , Independent Component Analysis (ICA) and with two other recently reported methods that use Deep Learning models (architecture used: Sparse Stacked Autoencoder NNs (SAE-NN) and Convolutional NN (CNN)) for the same data set. For the Fault Diagnosis problem, the proposed method is compared with Support Vector Machines (SVM), Random Forest, Structure SVM, and sm-NLPCA (architecture used: Stacked Autoencoder). It is observed that the proposed relevance based method significantly increases the average fault detection accuracy over other methods.

## References

[1] Piyush Agarwal and Hector Budman. Classification of profit-based operating regions for the tennessee eastman process using deep learning methods. *IFAC-PapersOnLine*, 52(1):556–561, 2019.

[2] Piyush Agarwal, Melih Tamer, M Hossein Sahraei, and Hector Budman. Deep learning for classification of profit-based operating regions in industrial processes. *Industrial & Engineering Chemistry Research*, 2019.

Table 1: Comparison of Fault Detection rate with different methods

| Fault | PCA (15 components) | | DPCA (22 components) | ICA (9 components) | | DL (2017) | DL (2019) | Proposed DL Methodology |
|---|---|---|---|---|---|---|---|---|
| | $T^2$ | SPE | $T^2$ | $I^2$ | AO | SAE-NN | CNN | DDSAE |
| 1 | 99.2 | 99.8 | 99 | 100 | 100 | 77.6 | 91.39 | 99.9 |
| 2 | 98 | 98.6 | 98 | 98 | 98 | 85 | 87.96 | 98.2 |
| 4 | 4.4 | 96.2 | 26 | 61 | 84 | 56.6 | 99.73 | 99.7 |
| 5 | 22.5 | 25.4 | 36 | 100 | 100 | 76 | 90.35 | 100 |
| 6 | 98.9 | 100 | 100 | 100 | 100 | 82.8 | 91.5 | 99.7 |
| 7 | 91.5 | 100 | 100 | 99 | 100 | 80.6 | 91.55 | 99.9 |
| 8 | 96.6 | 97.6 | 98 | 97 | 97 | 83 | 82.95 | 94 |
| 10 | 33.4 | 34.1 | 55 | 78 | 82 | 75.3 | 70.05 | 89.1 |
| 11 | 20.6 | 64.4 | 48 | 52 | 70 | 75.9 | 60.16 | 87.7 |
| 12 | 97.1 | 97.5 | 99 | 99 | 100 | 83.3 | 85.56 | 99.4 |
| 13 | 94 | 95.5 | 94 | 94 | 95 | 83.3 | 46.92 | 95.6 |
| 14 | 84.2 | 100 | 100 | 100 | 100 | 77.8 | 88.88 | 100 |
| 16 | 16.6 | 24.5 | 49 | 71 | 78 | 78.3 | 66.84 | 94 |
| 17 | 74.1 | 89.2 | 82 | 89 | 94 | 78 | 77.11 | 97.7 |
| 18 | 88.7 | 89.9 | 90 | 90 | 90 | 83.3 | 82.74 | 90.9 |
| 19 | 0.4 | 12.7 | 3 | 69 | 80 | 67.7 | 70.87 | 89.9 |
| 20 | 29.9 | 45 | 53 | 87 | 91 | 77.1 | 72.88 | 89.6 |
| Average | 61.77 | 74.72 | 72.35 | 87.29 | 91.70 | 77.74 | 79.84 | 95.60 |

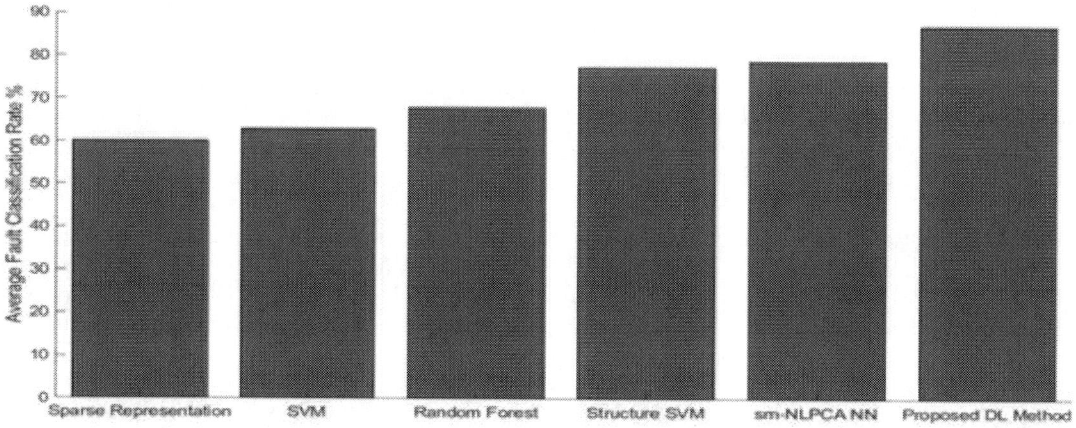


Figure 1: Comparison of Fault Classification rate with different methods

[3] Gavneet Singh Chadha and Andreas Schwung. Comparison of deep neural network architectures for fault detection in tennessee eastman process. In *2017 22nd IEEE International Conference on Emerging Technologies and Factory Automation (ETFA)*, pages 1–8. IEEE, 2017.

[4] Seongmin Heo and Jay H Lee. Statistical process monitoring of the tennessee eastman process using parallel autoassociative neural networks and a large dataset. *Processes*, 7(7):411, 2019.

[5] Barry M Wise, NL Ricker, DF Veltkamp, and Bruce R Kowalski. A theoretical basis for the use of principal component models for monitoring multivariate processes. *Process control and quality*, 1(1):41–51, 1990.

[6] Xiaofeng Yuan, Jiao Zhou, Yalin Wang, and Chunhua Yang. Multi-similarity measurement driven ensemble just-in-time learning for soft sensing of industrial processes. *Journal of Chemometrics*, 32(9):e3040, 2018.

# 601930 Physical Graph Neural Networks for Prediction of Fuel Ignition Quality

**Abstract:**

The increase in $CO_2$ emissions calls for the development of novel fuels that are based on renewable resources and can be used as a sustainable alternative in internal combustion engines (ICE). To assess the quality of a molecule as a fuel candidate, combustion-related properties need to be evaluated [1]. Fuel ignition quality indicators like the derived cetane number (DCN), the research octane number (RON), and the motor octane number (MON) give first, valuable insights on whether a molecule is suitable as an ICE fuel component. To predict these indicators and thus allow for facile molecular screening, many models based on quantitative structure-property relationships (QSPRs) have been proposed in the literature [2-5]. These models rely on manually selected descriptors to specify the molecular structure and apply regression methods to perform property prediction. The prediction quality, therefore, depends on the choice of informative descriptors.

Recently, graph neural networks (GNNs) [6,7] have shown promising results for the prediction of structure-property relationships of molecules [8]. GNNs use graph representations of molecules, where nodes correspond to atoms and edges to bonds. These graph representations are then linked to molecular properties by means of graph convolutions, pooling layers, and multilayer perceptron networks. More specifically, GNNs learn physico-chemical properties as a function of the molecular graph in a supervised learning set-up using a backpropagation algorithm. This end-to-end learning eliminates the need for selection of functional groups or descriptors. Rather, the method learns optimal molecular fingerprints through graph convolutions and maps the fingerprints to the physico-chemical properties by deep learning.

We propose a GNN model for predicting the DCN, MON, and RON of (oxygenated) hydrocarbons [9]. The model combines our higher order GNN with a recurrent neural network architecture. Further, it includes a physically meaningful pooling function which adds up contributions of individual atoms to the molecular descriptor. One of the main challenges in this work is the limited availability of experimental data. In order to overcome this difficulty, we study multitask learning, transfer learning, and ensemble learning. The results show competitive performance of the proposed GNN approach compared to state-of-the-art QSPRs making it a promising field for future research. The model and training code is open-source available at

https://git.rwth-aachen.de/avt.svt/public/graph_neural_network_for_fuel_ignition_quality
and a prediction tool is available via a web front-end at http://www.avt.rwth-aachen.de/gnn

[1] Gschwend, D., Soltic, P., Wokaun, A., & Vogel, F. (2019). Review and performance evaluation of fifty alternative liquid fuels for spark-ignition engines. Energy & Fuels, 33(3), 2186-2196.

[2] Dahmen, M., & Marquardt, W. (2015). A novel group contribution method for the prediction of the derived cetane number of oxygenated hydrocarbons. Energy & Fuels, 29(9), 5781-5801.

[3] Kubic Jr, W. L., Jenkins, R. W., Moore, C. M., Semelsberger, T. A., & Sutton, A. D. (2017). Artificial neural network based group contribution method for estimating cetane and octane numbers of hydrocarbons and oxygenated organic compounds. Industrial & Engineering Chemistry Research, 56(42), 12236-12245.

[4] Kessler, T., Sacia, E. R., Bell, A. T., & Mack, J. H. (2017). Artificial neural network based predictions of cetane number for furanic biofuel additives. Fuel, 206, 171-179.

[5] Smolenskii, E. A., Bavykin, V. M., Ryzhov, A. N., Slovokhotova, O. L., Chuvaeva, I. V., & Lapidus, A. L. (2008). Cetane numbers of hydrocarbons: calculations using optimal topological indices. Russian Chemical Bulletin, 57(3), 461-467.

[6] Kipf, T. N., & Welling, M. (2017). Semi-supervised classification with graph convolutional networks. Proceedings of the 5th International Conference on Learning Representations.

[7] Hamilton, W., Ying, R., & Leskovec J. (2017). Inductive Representation Learning on Large Graphs. Proceedings of the 30th Annual Conference on Neural Information Processing Systems, 1024-1034.

[8] Coley, C. W., Jin, W., Rogers, L., Jamison, T. F., Jaakkola, T. S., Green, W. H., Barzilay, R. & Jensen, K. F. (2019). A graph-convolutional neural network model for the prediction of chemical reactivity. Chemical Science, 10(2), 370-377.

[9] Schweidtmann, A. M., Rittig, J. G., König, A., Grohe, M., Mitsos, A., & Dahmen, M. (2020). Graph Neural Networks for Prediction of Fuel Ignition Quality. Submitted for Journal Publication.

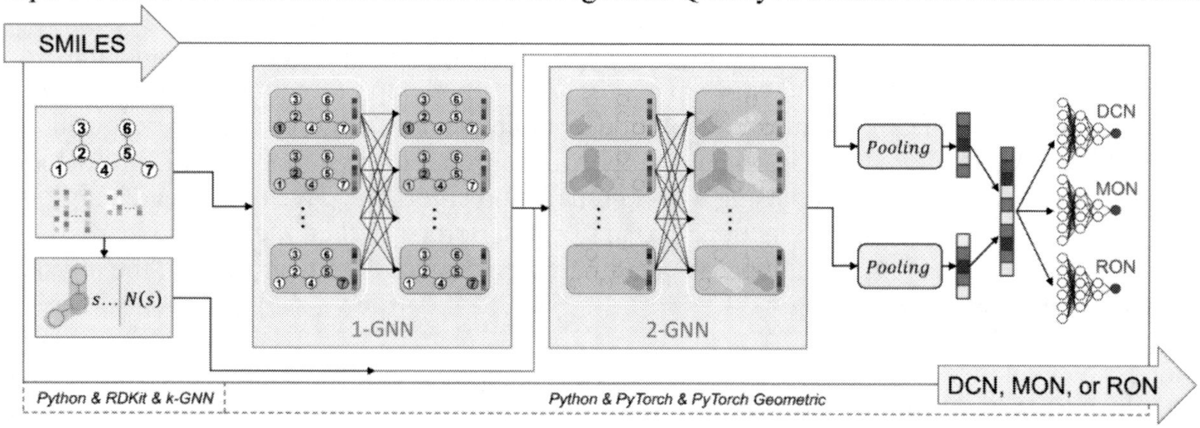

# 605089 Adaptive Spectral Graph Convolutional Neural Network in Crystal Property Prediction


**Abstract:**

Machine Learning (ML) has been more and more widely used in accelerating the design of new materials due to its close accuracy and less computational cost in predicting material properties comparing to simulation methods. However, existing ML methods for designing of crystalline materials are usually not flexible to handle different crystal types or hard to interpret because of the arbitrary size. Although there are more generalized models such as Crystal Graph Convolutional Neural Networks (CGCNN) which utilized graph CNN to encode the structure of a crystal, they are over-localized and impossible to learn beyond the bond connectivity. Here we show a stronger and more adaptive framework, based on spectral convolutional neural network, which is not only capable for arbitrary crystal structure but also learns a unique graph representation based on both global and local topology of each crystal.

First a crystal graph is produced base on the cif file of the crystal similar to CGCNN, and each node is encoded with its atomic properties such as electronegativity, covalent radius, etc. The connectivity is determined based on the distance between each atom. The graph Laplacian *L* is then obtained by performing graph Fourier transformation on the crystal graph. The transformation process can be decomposed by a complete set of eigenvectors U and the eigenvalue matrix LAMBDA($\lambda$), which represents the topology of the graph. The spectral filter $g_\theta(\lambda)$ generates a customized convolution kernel on the graph in space which can be formulated by

$g_\theta(\lambda)$=(sum from k=0 to k=k-1)$\sum \theta_k \lambda^k$

Where *k* can be interpreted as the number of convolutions in space.

However, this graph convolution cannot exploit the topological property of the graph. It is possible that the disconnected nodes have larger correlation than those connected nodes. To achieve the capability to learn from the geometric structure, we adapted a new spectral filter from Li et.al[1] which takes a non Euclean metric so that it learns a residual graph apart from the intrinsic crystal graph which makes a correction on the original graph Laplacian *L*:

L(with hat) = $L + \alpha L_{res}$

The overall intrinsic plus residue graph is trained with the spectral graph convoluted neural network to predict chemical properties of crystal and results are compared with various existing models.



[1] Li, R., Wang, S., Zhu, F. and Huang, J., 2018, April. Adaptive graph convolutional neural networks. In Thirty-second AAAI conference on artificial intelligence.

# 599718 Interpretable Model for Molecular Data Fusion

**Abstract:**

In this work, we introduce a graph-theoretical framework for fusing thermochemical data across different levels of quantum theory. We show that it can be used to predict *different thermochemical quantities at a higher level of theory using a quantity at a lower level of theory.* The generalizability of the model is investigated, and rigorous statistical tests are used to guarantee bounds in model predictions. Two important aspects of machine learnt models are addressed: domain knowledge and interpretability. We show how our model draws from chemical knowledge and thus provides an interpretable mapping across levels of theory and quantities of interest. These us to draw physical insight from the learning process. The approach is illustrated with multiple mapping tasks, and levels of quantum theory for a dataset of ~12k molecules. Chemical accuracy (1 kcal/mol) is approached for one task and is surpassed for all the others. Our approach provides a blueprint of how to merge disparate and incomplete literature datasets, built using different levels of theory, to create a more comprehensive thermochemical database for applications such as Big Data analysis.

# 601143 Leveraging Structure and Property Information for Building Maps of Materials

**Abstract:**

The application of machine learning techniques to problems in chemistry and materials science has become increasingly common over the past few years, with many studies focusing on property prediction or structural comparisons based on mathematical representations of materials. This has been made possible by the emergence of large, diverse databases of materials and molecules that allow for statistical analyses and high-throughput searches of candidate materials for a particular application. In many cases, such analyses are facilitated by the construction of a "map" that helps visualize the similarities and differences between the structures and properties of materials. Such maps are often based on a reduced structural representation obtained through the use of, for example, principal component analysis (PCA). However, maps constructed in this way do not necessarily reflect the property correlations between different structures. Similarly, the representation that yields the best property predictions may not be the best for highlighting structural relationships between different materials. In this work, we propose a kernel-based extension to Principal Covariates Regression (PCovR) [1], a method that combines PCA and linear regression, to create maps of a variety of materials, including molecular crystals and all-silica zeolites, that strike a balance between predictive ability and emphasizing structural differences. The kernelized extension, which we call Kernel Principal Covariates Regression (KPCovR), allows us to take advantage of nonlinear relationships in the data to build representations of materials that can offer improved regression performance over linear PCovR while preserving or enhancing the apparent structural diversity among the materials studied. Finally, we introduce a feature selection procedure based on PCovR aimed at identifying the features that best represent the structural diversity and predictive ability simultaneously, which makes it possible to achieve more accurate property predictions with a smaller number of features compared to other simple feature selection methods.

[1] S. de Jong, H. A. L. Kiers, *Chemomet. Intell. Lab. Sys. 14*, 155-164 (1992)

# 612158 Big Data-Based Fault Detection with Advanced Analytics in the Pharma Industry

**Abstract:**

Motivated by market pressure to reduce product costs, regulatory incentive to modernise manufacturing and a growing environmental concern to reduce waste has led the pharmaceutical industry to improve its manufacturing process. The impetus to adapt and modernise the manufacturing process can largely be attributed to the promotion of the Quality by Design paradigm (QbD) and guidance on Process Analytical Technology (PAT) by the FDA in 2004. The promotion of these methods coincided with a time where the cost of computing and data storage had declined, but also with developments in non-invasive online sensor technology, such as Raman and Near-Infrared (NIR) spectroscopy. This culminated into an increase intro the application of multivariate analysis (MVA) and specifically Multivariate Statistical Process Control (MSPC).

MSPC exploits the Big Data generated from a complex chemical process such as pharmaceutical production and can be used for process understanding, troubleshooting and on-line fault detection. MSPC tools include multivariant process charts and soft sensors and can be combined with APC methods to improve overall plant efficiency. MSPC aims to use big data to better understand a process, to develop a definition of normal operation and then use that to detect faults on-line. There are three main categories to classify as SPC methods (1) model-based (2) knowledge-based (3) data-based. Data-based methods are often the preferred method for pharmaceutical manufacturing. The advantages of data-driven methods are that they do not require first principals understanding of the process or a unit as they construct a model using input and output data. The most popular data-driven methods are multivariate latent/MVA methods (which include PCA, PLS and their adapted versions). These methods project data into the latent space and work as a feature extraction method and reduce the dataset into a smaller or manageable set that can be used in further applications of fault detection.

This paper presents an industrial case study which focused on fault detection towards operational improvements for a key API production, pursued in collaboration with a major multinational UK pharmaceutical company (GSK). The aim of this project is to exploit data availability towards constructing and validating a data-driven model of a specific API production unit of the GSK plant (Montrose, Scotland, UK). The project involves using large data sets of process data (temperatures, partial pressures, flow rates, and stream compositions, from an on-site GCMS analysis station), and implementing advanced MVA methods in MATLAB, in order to diagnose operational patterns and recommend feasible improvements towards reliable API production intensification.

LITERATURE REFERENCES

[1] E. Tomba et al., "General procedure to aid the development of continuous pharmaceutical processes using multivariate statistical modeling-An industrial case study," Int. J. Pharm., vol. 444, no. 1–2, pp. 25–39, 2013.

[2] T. Kourti, "Application of latent variable methods to process control and multivariate statistical process control in industry," Int. J. Adapt. Control Signal Process., vol. 19, no. 4, pp. 213–246, 2005.

[3] J. F. MacGregor and T. Kourti, "Statistical process control of multivariate processes," Control Eng. Pract., vol. 3, no. 3, pp. 403–414, 1995.

[4] T. Kourti, "Multivariate dynamic data modeling for analysis and statistical process control of batch processes, start-ups and grade transitions," J. Chemom., vol. 17, no. 1, pp. 93–109, 2003.

[5] J. F. MacGregor, P. Nomikos, and T. Kourti, "Multivariate Statistical Process Control of Batch Processes Using PCA and PLS," IFAC Proc. Vol., vol. 27, no. 2, pp. 523–528, 1994.

[6] S. Garcia-Munoz and D. Settell, "Application of multivariate latent variable modeling to pilot-scale spray drying monitoring and fault detection: Monitoring with fundamental knowledge," Comput. Chem. Eng., vol. 33, no. 12, pp. 2106–2110, 2009.

[7] M. Boiret, D. N. Rutledge, N. Gorretta, Y. M. Ginot, and J. M. Roger, "Application of independent component analysis on Raman images of a pharmaceutical drug product: Pure spectra determination and spatial distribution of constituents," J. Pharm. Biomed. Anal., vol. 90, pp. 78–84, 2014.

# 598331 Recent Advances in Kinetic Parameter Estimation Toolkit (KIPET) with Spectra

**Abstract:**

Developed over the past four years, the Kinetic Parameter Estimation Toolkit (KIPET), is an open-source toolbox for the determination of kinetic parameters from a variety of experimental datasets including spectra and concentrations. *KIPET* seeks to overcome limitations of standard parameter estimation packages by applying a unified optimization framework based on maximum likelihood principles and large-scale nonlinear programming strategies for solving estimation problems with nonlinear differential algebraic equations (DAEs). The software package includes tools for data preprocessing, determination of parameter confidence levels for a variety of problem types, and informative wavelength selection to improve the lack of fit (Schenk et al., 2020). All these features have been implemented in Python with the algebraic modeling package *Pyomo*. *KIPET* exploits the flexibility of *Pyomo to* formulate and discretize the dynamic optimization problems that arise in the parameter estimation algorithms. Solutions of these optimization problems are obtained with the nonlinear solver *IPOPT* and confidence intervals are obtained through the use of either *sIPOPT* or a newly developed tool, *k_aug*.

In this talk, we discuss a recently enhanced *KIPET* package (Short et al., 2020) that considers multiple experiments with potentially different reactants and kinetic models, different dataset sizes with shared or unshared individual species' spectra, leading to fast parameter estimation and confidence intervals based on the NLP sensitivities. In addition, we present a new variance estimation technique based on maximum likelihood derivations for unknown covariances from two sample populations. This approach leads to a straightforward deconvolution of variances between noise in model variables and in spectral measurements. Moreover, we discuss a new estimability analysis approach that systematically determines a ranked subset of kinetic parameters with well-defined confidence intervals (Chen and Biegler, 2020). With nonlinear kinetic models and limited measurements, it is often difficult to correctly estimate all the parameters, due to linear dependence and low correlation among the parameters. A common approach is to estimate a subset of the parameters by fixing the others at reasonable (often literature) values. However, it is often challenging to determine which parameters can be properly estimated. In this talk we present an efficient approach that ranks the estimable parameters, and discards those that cannot be estimated accurately. Based on reduced the reduced Hessian information with simple Gauss-Jordan elimination, the proposed approach leads to fast parameter selection and estimation within a simultaneous collocation framework. This approach becomes much more efficient for large problems than competing approaches based on multiple eigenvalue decompositions (Quaiser and Mönnigmann, 2009). Several case studies with increasing complexity are presented to demonstrate the performance of this proposed approach.

1. W. Chen, L. Biegler, "Reduced Hessian Based Parameter Selection and Estimation with Simultaneous Collocation Approach," submitted for publication (2020)
2. Schenk, M. Short, J. S. Rodriguez, D. Thierry, L. T. Biegler, S. Garcia-Munoz, W. Chen, "Introducing KIPET: A novel open-source software package for kinetic parameter estimation from experimental datasets including spectra," Computers and Chemical Engineering, 134(4), 106716 (2020)
3. Short, L. T. Biegler, S. Garcia-Munoz, W. Chen, "Estimating Variances and Kinetic Parameters from Spectra Across Multiple Datasets Using KIPET," Chemometrics and Intelligent Laboratory Systems, to appear (2020)
4. Quaiser, T., Mönnigmann, M. "Systematic identifiability testing for unambiguous mechanistic modeling-application to JAK-STAT, MAP kinase, and NF-κB signaling pathway models," BMC Systems Biology, 3,50 (2009).

# 601039 Teaching Computers to Interpret Mdsc Thermograms


**Abstract:**

We present a novel algorithm for the automatic data analysis of modulated Differential Scanning Calorimetry of thermograms designed to minimize subject matter expert input reduce the man reliability.

Differential Scanning Calorimetry (DSC) and mainly modulated DSC (mDSC) analysis are widely used to support the production of amorphous solid dispersions for the identification of kinetic and thermodynamic events such as glass transition, crystallization and melting. However, interpretation of thermograms require careful analysis of subject matter experts (SMEs) which can be time-consuming and thus can be a bottleneck to, e.g., formulation development where large numbers of samples are generated are require prompt analysis for reiteration. Furthermore, often there are elusive phenomena that can sometimes go unnoticed or cause dubious interpretation from different SMEs.

The presented Python-based software tool's computation of the derivatives and the maximums, minimums and inflections of each curve, is the base of the software. It then uses this information to search for events, based on their magnitude. One essential advantage of using an automated software is the ability to use the derivatives of the heat flow curves to improve the precision of the results. The derivatives are essential not only to the detect the exact temperature at which events occur but also to help define their onsets. In order to tackle the inherent noise of the DSC analysis and data acquisition, the proposed tool employs a low pass, Fourier transform-based filter, which enables the use of the thermogram derivatives.

We show case studies where the software analysis increases the detection rate of events that can easily be unnoticed by visual assessment due to their magnitude or shape, as they only cause very subtle variations. We also present case studies where smaller events are distorted or confounded with a nearby dominant events. However, following the digital detection of these events a more detailed visual assessment confirms their presence and often indicative of the onset of undesirable events such as phase separation or crystallization. Furthermore, the presented digital analysis tool promotes a faster and objective analysis of generated DSC thermogram data.


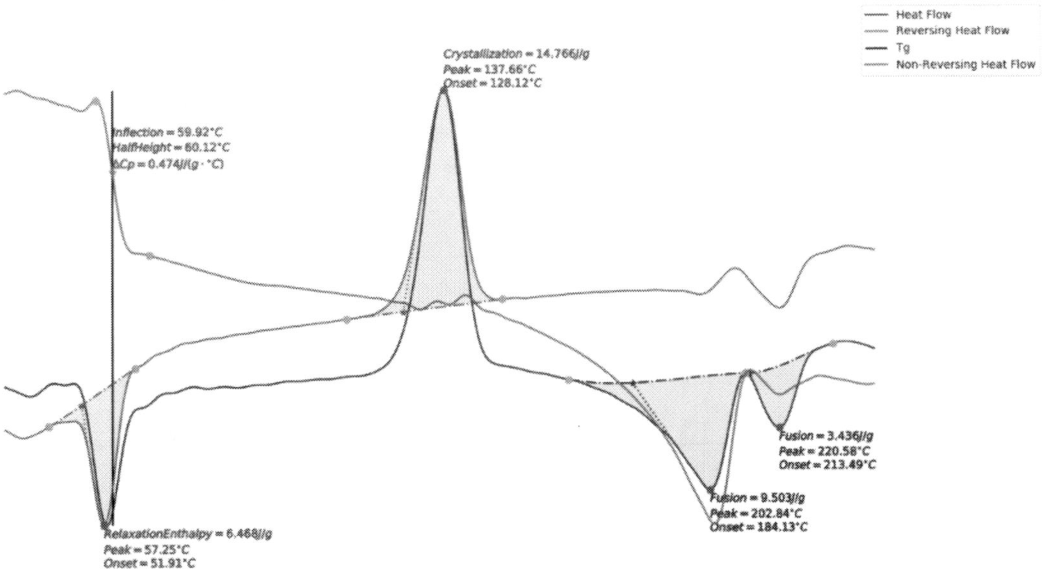

# 595673 Data-Smart Machine Learning Methods for Predicting Youngs Modulus of Directly Compressed Blends of Pharmaceutical Powders

**Abstract:**

The ability to predict the mechanical properties of compacted powder blends of Active Pharmaceutical Ingredients (API) and excipients solely from those of the individual components can reduce the amount of "trial-and-error" involved in formulation design. Machine Learning (ML) models can reduce model development time and effort with the imperative of adequate historical data. This work evaluates linear and non-linear ML for predicting the Youngs Modulus of directly compressed arbitrary blends of known excipients and API from readily available information about the API. The training data obtained from three BCS Class I APIs and four excipients demonstrate "data-smart" strategies to train ML models efficiently. The results indicate that even simple linear ML model provide reasonable prediction accuracy. The practical benefits of this method and how it compares with other mechanistic models are discussed. Finally, we demonstrate an application of the model to enable Quality-by-Design in drug product development.

# 593356 Advances and New Developments on Eiot for PAT Applications in Pharmaceutical Development and Manufacturing

**Abstract:**

The EIOT method[1] was proposed as an extension to the IOT[2] method as an alternative approach to extracting information from spectroscopic data into interpretable mass fractions of the resolved chemical species in the powder mixture.

The method in essence seeks to use Beer-Lambert's law as the deterministic functional relationship between spectra and mass fractions. The main difference with IOT is that EIOT does not assume the spectrum of the pure components is a measurable quantity; but a quantity that can only be estimated from measurements of known composition. The EIOT method proposes to decompose the measured spectra as the summation of the chemical signal plus the effect of the non-chemical interferences. The chemical signal is given by the summation of the apparent pure spectrum weighted by the mass fraction of each specie; the non-chemical interferences as the summation of the non-chemical interference signals weighed by their respective strength. The method relies on optimization techniques to estimate the apparent pure spectra and the signal of the non-chemical interferences; and when the method is used in real-time, to estimate the mass fractions and the strengths of the non-chemical interferences in each new sample.

Since the method was published, a number of enhancements have been done to the method, this talk presents such enhancements that i) the consideration of supervised signatures and ii) the consideration of exclusivity in a set of signatures (i.e. only one in a set is allowed to contribute). Details around the maintenance and long-term ownership of the EIOT method are discussed for its potential inclusion as a central component for a control strategy in drug product manufacture.

EIOT (as well as its predecessor IOT) bring a disruptive change in the practice of chemometrics; retiring the use of fixed regression models, to the use of real-time optimization and computer decision-making instead. As such, the requirements and expectations for use and long-term maintenance might have to be revised and adapted.

[1] Z. Shi, J. Hermiller, S.G.J.A.J. Muñoz, Estimation of mass-based composition in powder mixtures using Extended Iterative Optimization Technology (EIOT), (2018).

[2] K. Muteki, D.O. Blackwood, B. Maranzano, Y. Zhou, Y.A. Liu, K.R. Leeman, G.L. Reid, Mixture component prediction using iterative optimization technology (calibration-free/minimum approach), Industrial & Engineering Chemistry Research, 52 (2013) 12258-12268.

# 604116 A Data-Driven Approach to Understanding and Predicting the Early Formation of the Solid-Liquid Electrolyte Interphase

**Abstract:**

A remaining challenge in the pursuit of rational design of novel and optimized electrolytes is understanding and ultimately predicting the reaction cascade responsible for the creation of a functional liquid-anode solid-electrolyte interfaces (SEI). We here present an overview of the challenge, in different energy storage systems and advocate that a data-driven approach coupled with a high-throughput quantum chemistry computational infrastructure[1] can address the complexity of such reaction cascade. To this end, we have constructed a chemical reaction network containing over 50,000 elementary reactions which allow for bond breaking/formation as well as oxidation/reduction reactions. The reaction energetics are obtained from our computational framework which automate geometry optimization and vibrational frequency calculations for molecules, including radical, charged, metal-coordinated, and solvated species. Our framework, using select levels of theory and a suite of on-the-fly error handlers, is able to successfully converge to a minimum energy structure and calculate the molecular thermodynamic properties in over 97% of cases. Individual calculations can then be combined to perform complex workflows in a fully automated fashion. To date, we have applied this framework to over 25,000 unique molecules relevant to the formation of solid-electrolyte interphases in Li-ion and Mg-ion batteries. To analyze possible reaction pathways, we apply a graph representation to our reaction network, leveraging an iterative reweighting strategy that solves for all prerequisite branching such that traditional pathfinding algorithms correctly capture coordinated pathways. Machine-learning algorithms operating on bond-formation and breaking energetics aid in rapid evaluation of highly reactive processes. We show that without any apriori chemical intuition, our automated framework recovers the most favorable reaction paths to form key SEI components, which were carefully identified over the past two decades. Thus, our data-driven approach and infrastructure show promise to accelerate the understanding of chemical reactivity in complex environments, in the aid of novel electrolyte design.

# 595758 Accelerating the Discovery of Novel Perovskites with Robots and Data


**Abstract:**

In this talk, I will describe our Robotic-Accelerated Perovskite Investigation and Discovery (RAPID) system. The first generation of RAPID uses inverse temperature crystallization (ITC) to grow halide perovskite single crystals for x-ray structure determination and bulk characterization using commercial liquid handling robots. The second generation extends this to antisolvent vapor diffusion experiments. Experiment plans for the syntheses are contributed remotely, by both human scientists and algorithms trained on the reaction data, facilitating by our ESCALATE (Experiment Specification, Capture and Laboratory Automation Technology) data management system. Incoming data collected by ESCALATE is used to automatically train machine learning models, evaluate model performance and feature influence, and quantify reproducibility. A live web dashboard communicates these insights to the scientist and management in visual form. I will conclude by describing case studies about new scientific insights extracted from the comprehensive RAPID dataset that have been enabled by this comprehensive dataset, and discuss ongoing deployments of ESCALATE to perovskite thin film and vapor diffusion synthesis experiments.


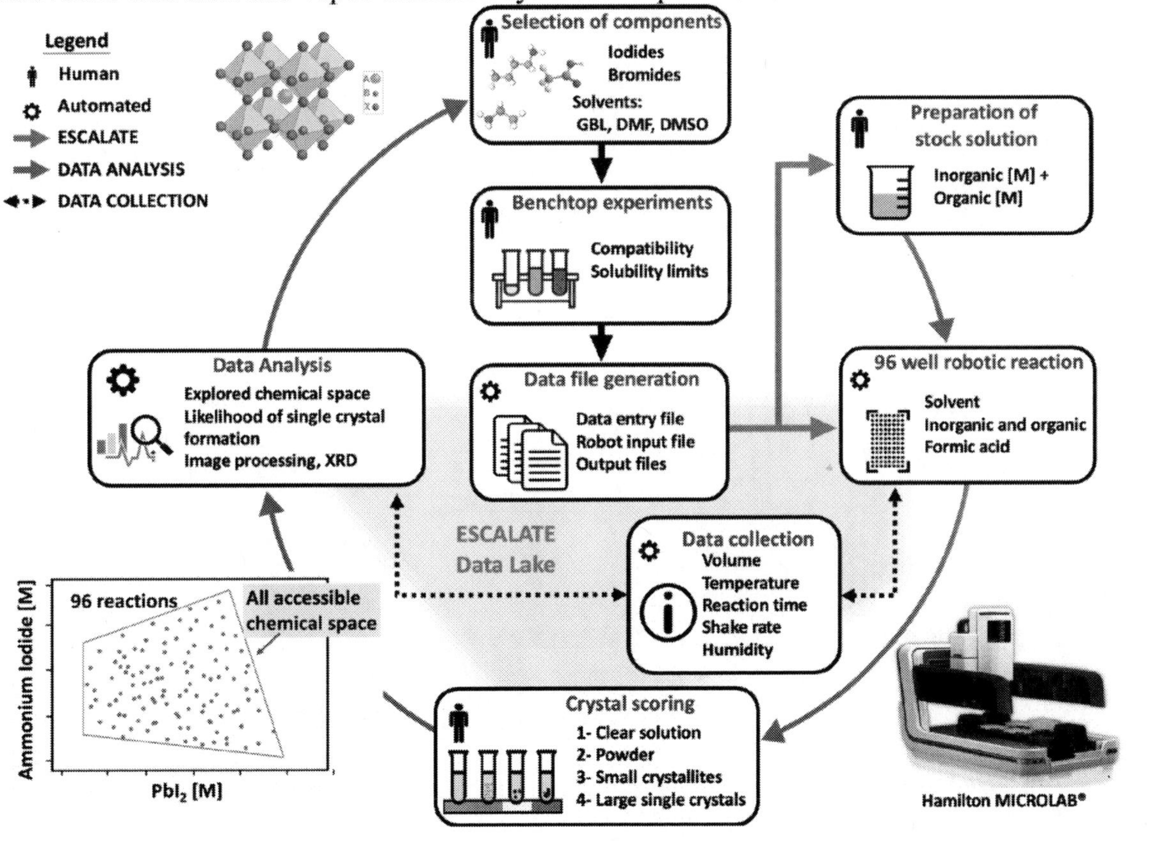


ESCALATE: Experiment Specification, Capture and Laboratory Automation Technology:
A software pipeline for automated chemical experimentation and data management

# 598796 Streamlining Small Molecule Design and Synthesis

**Abstract:**

This talk will cover recent progress in the application of data science and machine learning techniques to problems in synthetic chemistry as they relate to molecular discovery. The typical discovery paradigm is an iterative process of designing candidate compounds, synthesizing those compounds, and testing their performance, where each repeat of this cycle can require weeks or months. Time and cost constraints may necessitate selecting compounds on the basis of what are perceived as fast to synthesize, rather than what are most informative to assay. Data science and statistical learning offer unprecedented opportunities to systematize and streamline the process by which new functional small molecules are designed and synthesized. I will describe how historical reaction data can be used to inform decision-making in small molecule pathway design, for both retrosynthesis and forward reaction prediction, and how these techniques can be integrated with *de novo* molecular design to triage candidate compounds for testing.

# 609606 Transitioning from 'Traditional' to Data-Driven 'wet' Laboratories: Growing-Pains and Future Outlook

**Abstract:**

The adoption of data-science methods within chemical engineering is primed to revolutionize materials discovery and to accelerate developments in molecular understanding. Yet, the movement is largely dominated by data and information originating from molecular modelling and simulation, or that is acquired from legacy sources (e.g. databases, literature mining) accumulating data from decades of careful experimental work. In order to advance data-driven materials developments well into the future, high-throughput experimentation (HTE) and automation throughout the complete laboratory workflow must also be developed and widely adopted to accelerate rates of experimental data production. In this talk we outline several examples and experiences showcasing how our research group is developing and adapting hardware and software infrastructure to accelerate the pace of molecular discovery in soft-matter systems for applications in health care, clean energy and materials synthesis. The talk will highlight recent research examples related to the implementation of HTE for electrolyte discovery and colloidal formulation/synthesis. We will also highlight significant challenges that have emerged from transitioning an established 'wet-laboratory' practice to HTE. These relate to adapting routine experimental methods to achieve HTE, developing new skills within the research workforce, the adoption of new data stewardship practices, needs for autonomous data sorting/classification, algorithms for automatic modeling and analysis and many others. Conversely, we will also highlight the numerous opportunities that emerge for enhancing virtual collaboration, enabling open data/hardware/software sharing, tackling challenging irreducible problems (e.g. optimization of complex formulations), and improving the outlook for the implementation of 'self-driving' laboratories.

# 612393 Accelerating Molecular Discovery with Machine Learning and AI

**Abstract:**

Deep learning is revolutionizing many areas of science and technology, particularly in natural language processing, speech recognition and computer vision. In this talk, we will provide an overview into latest developments of machine learning and AI methods and application to the problem of drug discovery and development at Isayev's Lab at CMU. We identify several areas where existing methods have the potential to accelerate pharmaceutical research and disrupt more traditional approaches. First we will present a deep learning model that approximate solution of Schrodinger equation. Focusing on parametrization for drug-like organic molecules and proteins, we have developed a single 'universal' model which is highly accurate compared to reference quantum mechanical calculations at speeds 10^6 faster. Second, we proposed a novel computational method for *de-novo* design of molecular compounds with desired biological and physical properties.

# AUTHOR INDEX